ECOEFFICIENCY: THE BUSINESS LINK TO SUSTAINABLE DEVELOPMENT

ECOEFFICIENCY: THE BUSINESS LINK TO SUSTAINABLE DEVELOPMENT

Elisa Gomez Gonzalez

SOCIETY Publishing

www.societypublishing.com

Ecoefficiency: The Business Link to Sustainable Development

Elisa Gomez Gonzalez

Society Publishing
2010 Winston Park Drive,
2nd Floor
Oakville, ON L6H 5R7
Canada
www.societypublishing.com
Tel: 001-289-291-7705
001-905-616-2116
Fax: 001-289-291-7601
Email: orders@arclereducation.com

ISBN: 978-1-77407-364-3 (Hardcover)

ABOUT THE AUTHOR

Elisa Gomez was born and raised in Canary Islands; her personal mission is to incorporate Corporate Social Responsibility and Sustainability in all businesses. She strongly believes in the positive impact and added value it brings to the business itself and to all. With a Master's degree in International Corporate Social Responsibility and with more than 7 years of experience as Sustainability expert in International and Startups companies from a variety of sectors, Elisa is an expert in efficient and impactful Sustainability: Strategy definition, Implementation, Analysis, Training and/or Reporting.

TABLE OF CONTENTS

LIST OF FIGURES

LIST OF TABLES

LIST OF ABBREVIATIONS

AI	artificial intelligence
AT	attribution hypothesis
BCSD	Business Council for Sustainable Development
BICEP	Business for Innovative Climate and Energy Policy
CDSB	Climate Disclosure Standards Board
CESSA	Cemento de El Salvador S.A. de C.V.
CLG	Corporate Leadership Group
CSM	Communication and Social Marketing
CSR	Corporate Social Responsibility
CSR	corporate social obligation
EEA	European Environment Agency
EIB	European Investment Bank
EMS	environmental management system
ESG	ecological, social and administration
EVPs	employee value propositions
GRI	global reporting initiative
IDPs	individual development plans
IEA	International Energy Agency
IIGCC	Institutional Investors Group on Climate Change
INCR	investor network on climate risk
INDCs	intended nationally determined contributions
IoT	Internet of Things
IPCC	Intergovernmental Panel on Climate Change
KRAs	key result area
LTIs	long-term incentive compensation
MDPs	management development programs

MENA	Middle East and North Africa
MSF	multi-stakeholder feedback
NDCs	nationally determined contributions
NGOs	non-legislative businesses
OECD	Organization for Economic Co-operation and Development
RFP	request for proposals
RPET	reused polyethylene terephthalate
SCM	supply chain management
SDGs	sustainable development goals
SRB	sustainable mindful business
SS/HS	safe schools/healthy students
SSA	sub-Saharan Africa
TEPCO	Tokyo Electric Power Company
TPP	trans-pacific partnership
TTIP	Transatlantic Trade and Investment Partnership
UN	United Nations
UNEP	United Nations Environment Program
WBCSD	World Business Council for Sustainable Development
WSSD	World Summit on Sustainable Development
WTO	World Trade Organization

SYNOPSIS

This book will run you through the entire process of setting up a sustainability management system, that everybody feels accountable and responsible for.

The first chapters will set up the scene describing how sustainability is currently being implemented, what are the trends and challenges. These initial chapters are very important as they contemplate the main assumptions and characteristics of current practices, which will enable the reader to better understand the measures presented afterwards.

Chapter 7 goes in detail towards the trends on renewable technology, the future 10 years of development, and the different scenarios and challenges encountered in all regions of the world.

Chapter 10 goes in details towards explaining what to aim for and how to implement actions bearing in mind the essentials of sustainability practices amongst stakeholders.

Chapter 11 covers compliance with best practices and international standards, which normally tend to be a great part in any given marketing strategy.

The last chapters focus mainly on the above four main areas where sustainability practices and specifically eco-efficiency and eco-innovation have the greatest inputs.

EXECUTIVE SUMMARY

An Expanded Concentration on Sustainability Matters

Sustainable business examples are progressively showing up on the radar of organizations around the globe because of current trends and communication efforts. An effective company or association is very much reliant on its investor's needs and expectation. Investors are interested in sustainability, impact, and added value.

A reasonable association meets the desires of diverse partners. A genuine feasible association is capable of bringing together monetary, social, and natural contemplations. Such businesses are effectively ready to meet the desires of both inward and outer partners, for example, investors, clients, government, employees, and groups to guarantee long-term success for all.

In any case, most businesses neglect to execute feasible objectives. Regularly, businesses neglect to execute sustainable business practices because of insufficient reconciliation with business methodology, absence of help from authority, constrained comprehension and inclusion of workers at all levels and additionally nonappearance of frameworks and procedures that enable a systematize support.

Sustainable can enable businesses to prevail with regards to current economical business trends. Albeit numerous businesses dedicate themselves to sustainability and frame a strategy, its implementation and coordination depend upon effectively embracing the following:

- Empowering and driving economical business practices at all levels of the association.
- Creating effective frameworks and procedures.

INTRODUCTION

Sustainability emerged from the need to protect and ensure our environment, our society, and our economic system assumes responsibility and works to cause as little harm as possible. Sustainability as a concept stepped onto the world stage in 1992 at the World's first Earth Summit in Rio. Since then, awareness and the application of sustainable practices have been on the rise.

This book will include the background information needed to understand sustainability management systems and their integration into businesses fully. The reader will be provided with the benefits, risks, and disadvantages of the different methods to assure comprehension and analytical thinking is stimulated.

Renewable technology development goes hand-in-hand with sustainability management systems. Thus, this book covers the trends, future expectations, and forecasts as well as a thorough overview of what is happening around the world.

A simple methodology will be presented to effectively approach the integration of sustainability in all types of businesses. Case studies, practical examples, and templates will be provided in each phase of the methodology, enabling the reader to assess and implement sustainability as they read along. Relevant supporting information will be provided at the end of each chapter on specific topics to take into consideration.

Promoting and empowering people into incorporating sustainability are my personal mission. I strongly believe in the positive impact it provides and in the added value that it enables for all. Sustainability does not have to be complicated, expensive, or unrelated to the business activity. The numerous benefits its incorporation provides to all types of companies are always worth the journey.

In the end, the main goal of sustainability, in my humble opinion, is to do and use business as a force for good.

PREFACE

Sustainability management systems have proven not only to benefit the company that applies them but also the world as a whole. Sustainability is slowly but steadily being incorporated into public policies, governments, private sector companies, and even in freelancers. The world has begun to believe that we can actually operate and perform in a way which is respectful with the environment, society, and the economy enabling us to develop in a sustainable manner.

This is not merely a trend or the hot topic for the coming years, but it is a crucial aspect of we want to continue prospering a thriving. A great part of sustainable management practices relies on people. People that work with a purpose; feel engaged, accountable and responsible for their actions. People that want to make a difference for good. These people together form teams that work with purpose, feel engaged, connected, accountable and responsible for their actions. These group of people forms companies, businesses, organizations and overall a community of people that have in mind to be sustainable and to provide the best for future generations.

The purpose is the main driver for millennials in choosing a workplace and a career, this shift from other generational trends implies that companies and the workplace industry globally need to face a transformation. Millennials which will shortly constitute the biggest generation in the workplace have a broader understanding of sustainability. Not only a broader understanding but they go one step further and actually reject working for companies that do not create value and have a positive impact where they operate. Their understanding of sustainable development together with the immense technological capacity nowadays set the perfect scene for all types of companies and organizations to start implementing sustainability management systems.

With sustainability management systems, the renewable technology development goes hand-in-hand. Thus, along with an overview of what is happening around the world, this book covers the trends, future expectations, and forecasts .

Typically sustainability management systems or sometimes referred to as Corporate Social Responsibility (CSR) is very much dependent upon international standards, regulation, legislation, and certifications which are very much industry-related or country-related. This means that they are heavily influenced by the culture, the economic positioning in the market and the countries legal and political trends.

Moreover, globalization has enabled access to products and services where they were not available before. All of these changes and limitations have created a difficult scenario for companies, where implementing sustainability measures implies allocating a great part of the budget and result in very time consuming for employees.

The main aim of this book is to provide simple, effective, understandable and applicable mechanisms for all readers. To enable and empower them to implement management systems that is adapted to the reality and expectations of their business.

In this book, you will find all of those mechanisms and resources needed to ensure that sustainability is well embedded into a company through Human Resources Practices, Marketing and Communication, Procurement, and CEO Leadership.

This book is for all those that want to effectively implement sustainable management and culture into their current business practices. The book runs simply through all of the steps to design and create a strategy according to the needs and expectations of all the stakeholders ensuring everyone involved feels accountable and responsible for their piece of the pie.

It is also intended for all those studying or professionals that work as sustainability consultants or analysts to broaden their perceptions of sustainability implementation in a millennial digital world.

CHAPTER 1

What Is Sustainability?

There is no specific definition for sustainability; over the years, the concept has changed and can be modified to meet the needs and requirements of the different agents and where we are applying sustainability. Institutions, companies, governments, and individual experts all have their own interpretation of what sustainability is for their specific area of influence or perspective in which they view the world.

In order to have a further insight of what sustainability can be defined as you have here a selection of different definitions:

- **The World Commission on Environment and Development:** A process of change in which the exploitation of resources, the direction of investments, the orientation of technological development and institutional change are all in harmony and enhance both current and future potential to meet human needs and aspirations.
- **Forum for the Future:** Sustainable development is a dynamic process which enables people to realize their potential and improve their quality of life in ways which simultaneously protect and enhance the earth's life support systems.
- **Future's Sustainable Wealth London Project:** In essence, sustainable development is about five key principles: quality of life; fairness and equity; participation and partnership; care for our environment and respect for ecological constraints – recognizing there are 'environmental limits'; and thought for the future and the precautionary principle.
- **The Real-World Coalition 1996, a Definition Based on**

the Work of the World Commission on Environment and Development: The environment must be protected. To preserve essential ecosystem functions and to provide for the wellbeing of future generations; environmental and economic policy must be integrated; the goal of policy should be an improvement in the overall quality of life, not just income growth; poverty must be ended and resources distributed more equally; and all sections of society must be involved in decision making.

- **The Dorset Education for Sustainability Network:** We cannot just add sustainable development to our current list of things to do but must learn to integrate the concepts into everything that we do.
- **Learning for a Sustainable Future – Teacher Centre:** A sustainable future is one in which a healthy environment, economic prosperity, and social justice are pursued simultaneously to ensure the wellbeing and quality of life of present and future generations. Education is crucial to attaining that future.
- **The Shaky Ground of Sustainable Development (Donald Worster in Global Ecology, 1993):** The first and perhaps most difficult problem, one that seldom gets addressed, is the time frame is a sustainable society one that endures for a decade, a human lifetime, or a thousand years? As you can see, there are various kinds of definitions for sustainability. Over the years, the definitions have evolved to better adapt to the current challenges and or based on the organization that addresses a given cause. For the purpose of better understanding, the content of this book and the proposed solutions to each marketing aspect here presented, we will base on the definitions of sustainability presented in the next chapter.

CHAPTER 2

Sustainable Development

According to the Brundtland Report, *Sustainable Development* is the development that meets the needs of the present without compromising the ability of future generations to meet their own needs. It contains within it two key concepts:

- The concept of needs, in particular, the essential needs of the world's poor, to which overriding priority should be given; and
- The idea of limitations imposed by the state of technology and social organization on the environment's ability to meet present and future needs.

2.1. SUSTAINABILITY

The ability to be sustained, supported, upheld, or confirmed. The quality of not being harmful to the environment, economy, society or depleting natural resources, and thereby supporting long-term ecological balance.

2.2. CORPORATE SOCIAL RESPONSIBILITY (CSR)

The movement aimed at encouraging companies to be more aware of the impact of their business on the rest of society, including their own stakeholders and the environment. CSR is a business approach that contributes to sustainable development by delivering economic, social and environmental benefits for all stakeholders. CSR is a concept with many definitions and practices. The way it is understood and implemented differs greatly for each company and country. Moreover, CSR is a very broad concept that addresses many and various topics such as human rights, corporate governance, health and safety, environmental effects, working conditions and contribution to economic development. Whatever the definition is, the purpose of CSR is to drive change towards sustainability. Although some companies may achieve remarkable efforts with unique CSR initiatives, it is difficult to be on the forefront on all aspects of CSR. Considering this, the example below provides good practices on one aspect of CSR – environmental sustainability.

2.2.1. Example

Unilever is a multinational corporation, in the food and beverage sector, with a comprehensive CSR strategy. The company has been ranked 'Food Industry leader' in the Dow Jones Sustainability World Indexes for the 11 consecutive years and ranked 7th in the 'Global 100 Most Sustainable Corporations in the World.' One of the major and unique initiatives is the 'sustainable tea' program. On a partnership-based model with the Rainforest Alliance (an NGO), Unilever aims to source all of its Lipton and PG Tips tea bags from Rainforest Alliance Certified™ farms by 2015. The Rainforest Alliance Certification offers farms a way to differentiate their products as being socially, economically and environmentally sustainable.

CHAPTER 3

Why Is It Important?

"Creating a strong business and building a better world are not conflicting goals – they are both essential ingredients for long-term success."
– *William Clay Ford, Jr. Executive Chairman, Ford Motor Company*

Having a sense of sustainability and successfully incorporating CSR into business practices and management brings along a series of short-term and long-term benefits. In this section, we will go through the top nine benefits that CSR practices can provide to any company.

1. **Better Risk Management and Ability to Address Change:** CSR will help you better anticipate and manage the wide range of risks that your company is confronted with. Risks are generally very dependent on your type of activity and sector; in this case, it goes beyond. Having a good understanding of your impacts as a company will highlight *the* risks you are taking, making it easier for you to address them. Having well identified your risks will enhance your ability to address change.

"In the next decade, the most successful companies will be those that integrate sustainability into their core businesses."
– *Jim Owens, CEO Caterpillar*

2. **Improved Reputation Management:** You are working with a purpose and a well-established set of values. You are who you are and you work for it constantly and coherently every

day. This makes you trustworthy. When your people (customers, suppliers, investors, community, etc.) feel they can trust you, they will stay with you. Long-term relations are the best way to measure your reputation.

"To build and sustain brands people love and trust, one must focus – not only on today but also on tomorrow. It's not easy, but balancing the short and long-term is key to delivering sustainable, profitable growth – growth that is good for our shareholders but also good for our consumers, our employees, our business partners, the communities where we live and work, and the planet we inhabit."

– Irene B. Rosenfeld, CEO Kraft

3. **Satisfied Employees:** One of the main business benefits of sustainability is higher engagement levels from your employees. Better understanding of what they want, what they expect and why they will stay. Your staff is your greatest asset; keeping them motivated and trained will make the best version of your company come to life.

"The only thing worse than training employees and losing them is to not train them and keep them."

– Zig Ziglar

4. **Access to Innovation, Competitiveness and Market Positioning:** By having open dialogue with your stakeholders you will have a better understanding of their needs and expectations. Knowing what they know and expect will give you the advantage of access to information. With this knowledge, you can better adapt and improve in order to earn higher competitiveness and market positioning. In order to innovate, you have to have a great team and a lot of valuable information.
5. **Operational Efficiency and Cost Savings:** Revising your production channels to assure you are operating in the least impactful way according to law and best practices will not only have a positive impact in your cost savings but also with your future activity. Operating sustainability will guarantee you stay in business.

6. **Efficient Supply Chains:** CSR will help you improve your ability to both attract and develop effective and efficient relationships with your suppliers. Without them, your business would probably not work, and the way they deliver has an impact in the way you will deliver. Open communication and take action to evaluate their performance.
7. **Social License:** Addressing the community where you operate and getting to know up to what extent you affect or influence in them will give you a more robust social license. Social license generally refers to a local community's acceptance or approval of a company's project or ongoing presence in an area. It is increasingly recognized by various stakeholders and communities as a prerequisite to development.

"We know that the profitable growth of our company depends on the economic, environmental, and social sustainability of our communities across the world. And we know it is in our best interests to contribute to the sustainability of those communities"

– ***Travis Engen, CEO, Alcan***

8. **Access to Capital and Investors:** Investors are more and more interested in companies that work for a clear purpose that will help the world to be a better place, companies that are responsible with society and the environment. Check out FTSE4good index series to have a view of what is available for companies.
9. **Your Contribution to a Better World:** By implementing CSR you are assuming your part of responsibility and taking action, you are contributing to international issues and taking a side on them.

"The business of business should not be about money. It should be about responsibility. It should be about public good, not private greed"

– ***Anita Roddick, Founder, Body Shop***

CHAPTER 4

The Evolution of CSR Practices

"Our biggest challenge this century is to take an idea that seems abstract – Sustainable Development – and turn it into a reality for the world's peoples."

– Kofi Annan UN Secretary-General 1997–2007

4.1. THE PYRAMID OF CSR

The four-section meaning of CSR was initially distributed in 1979. In 1991, Carroll extricated the four-section definition and recast it as a CSR pyramid. The reason for the pyramid was to single out the definitional part of CSR and to show the building piece nature of the four-section structure. The pyramid was chosen as a geometric outline since it is straightforward, instinctive, and worked to withstand the trial of time.

Subsequently, the financial duty was set as the base of the pyramid since it is a foundational prerequisite in business. Similarly as the footings of a building must be solid to bolster the whole structure, managed benefit must be solid to bolster society's different desires of undertakings. The point here is that the foundation of CSR is based upon the introduction of a financially solid and supportable business.

In the meantime, society is passing on the message to business that it is required to comply with the law and consent to controls since law and directions are society's codification of the essential guidelines whereupon business is to work in a common society. On the off chance that one takes a gander at CSR in creating nations, for instance, regardless of whether a lawful and administrative structure exists or not altogether influences whether multinationals contribute there or not on the grounds that such a lawful framework is basic to give an establishment to honest to goodness business development.

What's more, business is relied upon to work in a moral form. This implies business has the desire, and commitment, that it will make the right decision, just, and reasonable and to maintain a strategic distance from or limit damage to every one of the partners with whom it collaborates. At last, business is relied upon to be a decent corporate national, that is, to give back and to contribute budgetary, physical, and HR to the groups of which it is a section. So, the pyramid is inherent a design that mirrors the crucial parts played and expected by business in the public arena. Figure 4.1 shows a graphical delineation of Carroll's Pyramid of CSR.

Figure 4.1: Carroll's pyramid of CSR.

Source: Schwartz and Carroll (2003)

4.2. ETHICS PERMEATES THE PYRAMID

In spite of the fact that the moral obligation is portrayed in the pyramid as a different classification of CSR, it ought to likewise be viewed as a component which slices through and immerses the whole pyramid.

Moral contemplations are available in each of the other duty classifications also. In the Economic Responsibility classification, for instance, the pyramid certainly accepts a free enterprise society wherein the mission for benefits is seen as a real, just desire. Free enterprise, as it were, is a monetary framework which considers it being morally proper that proprietors or shareholders justify an arrival on their ventures.

In the Legal Responsibility class, it ought to be recognized that most laws and directions were made in light of some moral thinking that they were suitable. Most laws became out of moral issues, e.g., a worry for shopper security, worker wellbeing, the common habitat, and so on, and along these lines once formalized they spoke to "classified morals" for that society. What's more, obviously, the Ethical Responsibility remains all alone in the four section shown as a classification that grasps strategies and practices that many see as living at a larger amount of desire than the essentials required by law. Insignificantly, law may be viewed as inactive consistency. Morals, on the other hand, recommend a level of lead that may suspect future laws and in any occasion, endeavor to do what is considered above most laws, that which is driven by integrity.

At long last, Philanthropic Responsibilities are in some cases morally persuaded by organizations endeavoring to make the best choice. Despite the fact that a few organizations seek after magnanimous exercises as a utilitarian choice (e.g., key generosity) just to be viewed as "great corporate residents," some do seek after charity since they view it as the temperate thing to do. In this last understanding, magnanimity supposedly is morally inspired or charitable in nature (Schwartz and Carroll, 2003). In rundown, moral inspirations and issues sliced through and penetrate every one of the four of the CSR classifications and along these lines accept a crucial part in the totality of CSR.

4.3. TENSIONS AND TRADE-OFFS

As organizations look to enough perform as for their monetary, legitimate, moral and humanitarian duties, pressures and exchange offs definitely emerge. How organizations choose to adjust these different obligations goes far towards characterizing their CSR introduction and notoriety.

The monetary obligation to proprietors or shareholders requires a watchful exchange off between here and now and long haul productivity. In the short run, organizations' consumptions on legitimate, moral and humanitarian commitments perpetually will "show up" to struggle with their duties to their shareholders. As organizations use assets on these obligations that have the earmarks of being in the essential interests of different partners, a test to remove corners or look for best long range favorable circumstances emerges. This is when strains and exchange offs emerging. The customary believed is that assets spent for legitimate, moral and generous purposes may fundamentally reduce productivity.

Be that as it may, as indicated by the "business case" for CSR, this is not a substantial presumption or conclusion. For quite a while it has been the developing perspective that social action can and leads to financial prizes and that business ought to endeavor to make such a positive circumstance (Chrisman and Carroll, 1984).

The business case for CSR alludes to the fundamental contentions supporting or reporting why the business group ought to acknowledge and propel the CSR cause.

The business case is worried with the essential question – What does the business group and business ventures escape CSR? That is, how would they profit unmistakably and specifically from taking part in CSR approaches, exercises and practices (Carroll and Shabana, 2010).

There are numerous business case contentions that have been made in the writing; however, four successful contentions have been made by Kurucz et al., and these incorporate cost and hazard decreases, beneficial outcomes on upper hand, organization authenticity and notoriety, and the part of CSR in making win-win circumstances for the organization and society (Kurucz et al., 2008).

Different reviews have counted the explanations behind business to grasp CSR to incorporate advancement, mark separation, worker engagement, and client engagement. The reason for business case thinking regarding the Pyramid of CSR is to improve the trusted clashes and pressures between and among the four classifications of duties. To put it plainly, the pressures and tradeoffs will keep on being vital choice focuses; however, they are not in total resistance to each other as is frequently seen.

4.4. THE INTEGRATED PYRAMID OF CSR

The Pyramid of CSR is planned to be seen from a partner viewpoint wherein the emphasis is all in all, not the diverse parts. The CSR pyramid holds that organizations ought to participate in choices, activities, arrangements, and practices that the while satisfies the four segment parts. The pyramid ought not to be deciphered to imply that business is relied upon to satisfy its social duties in some consecutive, various leveled, mold, beginning at the base.

Or maybe, business is relied upon to satisfy all duties at the same time. The situating or requesting of the four classifications of obligation endeavors to depict the essential or fundamental nature of these four classifications to business' presence in the public arena. As said sometime recently, monetary and legitimate duties are required; moral and charitable obligations are normal and craved. The portrayal being depicted, accordingly, is that the aggregate social duty of business involves the simultaneous satisfaction of the association's monetary, lawful, moral, and altruistic duties.

Expressed as a condition, it would read as takes after: Economic Responsibilities + Legal responsibilities + Ethical Responsibilities + Philanthropic Responsibilities = Total Corporate Social Responsibility (CSR).

Expressed in more functional and administrative terms, the CSR driven firm ought to endeavor to make a benefit, comply with the law, take part in moral practices and be a decent corporate native. At the point when found along these lines, the pyramid is seen as a brought together or coordinated entire (Carroll and Buchholtz, 2015).

4.5. THE PYRAMID IS A SUSTAINABLE STAKEHOLDER FRAMEWORK

Each of the four segments of duty locations diverse partners regarding the changing needs in which the partners may be influenced. Monetary duties most drastically affect shareholders and representatives on the grounds that if the business is not monetarily reasonable both of these gatherings will be essentially influenced.

Legitimate obligations are positively essential as for proprietors; however, in today's hostile society, the risk of prosecution against organizations emerge regularly from workers and shopper partners.

Moral duties influence all partner gatherings. Shareholder claims are an extending class. At the point when an examination of the moral issues business confronts today is considered, they regularly include workers, clients, and the earth generally every now and again.

At last, magnanimous duties most influence the group and charitable associations; additionally, workers since some examination has inferred that an organization's humanitarian inclusion is essentially identified with its representatives' resolve and engagement. The pyramid ought to be viewed as practical in that these duties speak to long haul commitments that overarch into future eras of partners also. In spite of the fact that the pyramid could be seen to be a static preview of duties, it is proposed to be viewed as a dynamic, versatile system the substance of which spotlights both on the present and what's to come.

A thought of partners and supportability, today, is indivisible from CSR. In fact, there have been a few interests in the writing for CSR to be reclassified as Corporate Stakeholder Responsibility and others have supported Corporate Sustainability Responsibilities. These interests highlight the personal way of these interrelated themes (Carroll and Buchholtz, 2015).

Moreover, Ethical Corporation Magazine which underscores CSR in its Responsible Summit meetings coordinates these two subjects – CSR and Sustainability – as though they were one and, truth be told, numerous business associations today see them along these lines; that is, to be socially dependable is to put resources into the significance of supportability which verifiably is worried with what's to come. Yearly corporate social execution reports regularly pass by the titles of CSR and additionally Sustainability Reports yet their substance are undifferentiated from each other; at the end of the day, the ideas are being utilized reciprocally by numerous.

4.6. GLOBAL APPLICABILITY AND DIFFERENT CONTEXTS

At the point when Carroll built up his unique four-section develops of CSR (1979) and after that his pyramidal delineation of CSR (1991), it was unmistakably finished in view of American-sort industrialist social orders. Around then, CSR was most common in these free endeavor social orders. Since that time, a few authors have recommended that the pyramid should be reordered to meet the states of different nations or littler organizations.

In 2007, Crane and Matten watched that every one of the levels of CSR portrayed in Carroll's pyramid assume a part in Europe yet they have a divergent centrality and are interlinked in a to some degree distinctive way (Crane and Matten, 2007). Moreover, Visser returned to Carroll's pyramid in creating nations/mainlands, specifically, Africa, and contended that the request of the CSR layers there vary from the exemplary pyramid. He goes ahead to state that in creating nations, financial duty keeps on getting the most accentuation; however, magnanimity is given second most elevated need taken after by legitimate and afterward moral duties (Visser, 2011).

Visser keeps on battling that there are myths about CSR in creating nations and that one of them is that "CSR is the same the world over." Following this, he keeps up that every locale; nation or group has an alternate arrangement of drivers of CSR. Among the "glocal" (global + local) drivers of CSR, he recommends that social convention, political change, financial needs, administration crevices, and emergency reaction are among the most vital (Visser, 2011, p. 269). Crane, Matten, and Spence make a decent showing with regards to talking about CSR in a worldwide setting when they expound on CSR in various locales of the globe, CSR in created nations, CSR in creating nations, and CSR in rising/transitional economies (Crane et al., (2008)).

Notwithstanding issues being raised about the appropriateness of CSR and, along these lines, the CSR pyramid in various territories, the same might be said for its pertinence in various authoritative settings. Settings of enthusiasm here might incorporate private part (substantial versus little firms), open part, and common society associations (Crane et al., (2008)). In one specific hypothetical article, Laura Spence looked to reframe Carroll's CSR pyramid, improving its significance for private venture. Spence utilized the ethic of care and women's activist viewpoints to redraw the four CSR areas by demonstrating that Carroll's classes spoken to a masculinist point of view however that the ethic of care viewpoint would concentrate on various concerns. In this way, she contended that the monetary obligation would be viewed as "survival" in the ethics of care viewpoint; legitimate would be viewed as "survival;" moral would be recast as ethics of care; and charity would keep on being altruism.

It may be watched that these are not totally contradictory with Carroll's classes. She then included another classification and that would be distinguished as "individual uprightness." She recommended that there could be no less than four private companies social duty pyramids – to self and family; to workers; to the neighborhood group; and to business accomplices (Spence, 2016). Certainly, different analysts will keep on exploring the appropriateness of the Pyramid of CSR to various worldwide, situational, and hierarchical settings. This is the manner by which hypothesis and practice create (Figures 4.2 and 4.3).

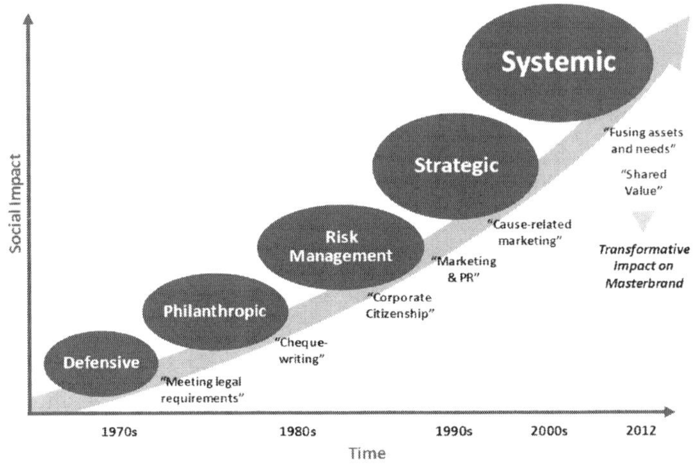

Figure 4.2: Evolving trends in corporate social responsibility from a strategy perspective.

Source: Crane et al., (2008)

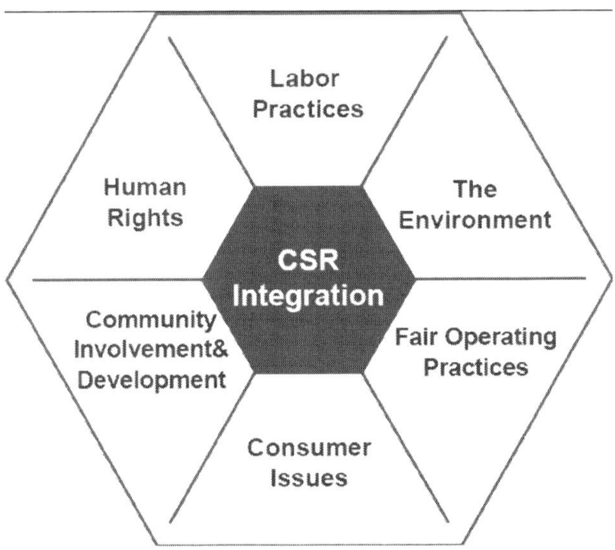

Figure 4.3: Evolving trends in corporate social responsibility from a management perspective.

Source: Crane et al., (2008)

CHAPTER 5

Sustainable Development Model Rethinking the Concept of Brand Value

- **Shareholder View:** Activities focused solely on maximizing short-term profits.
- **Stakeholder View:** Activities designed to maximize returns for employees and partners.
- **Societal View:** Activities designed to contribute to the greater good for all (Figure 5.1).

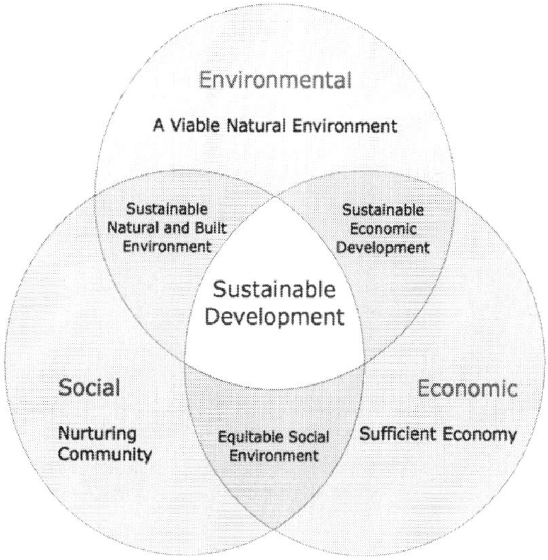

Figure 5.1: Corporate social responsibility model according to Sustainable Development.

Source: Crane et al., (2008)

CHAPTER 6

Common Excuses to Not Implement CSR

- CSR isn't for private ventures (just substantial partnerships). CSR works for vast and little associations alike.
- Don't attempt to beat the framework or swindle. Take an alternate way or manufacture results and anybody discovers – your notoriety gets destroyed and there is no easy route out of that.
- Don't invest the least exertion and call it CSR. Reusing a couple of boxes doesn't make you carbon nonpartisan. It's alright, to begin with little changes – simply report their outcomes sincerely and develop from that point.
- Don't just give cash away without an arrangement. CSR does not equivalent charity. It's a long-haul system, not a one-time money related gift.
- Don't invest so much energy composing an arrangement, that nobody gets an opportunity to tail it. Detail some straightforward procedures and execute them in an auspicious way.
- CSR is excessively convoluted. It doesn't need to be.
- Don't expect a direct money related ROI on CSR. A few returns are not effortlessly quantifiable, and as a long- haul system, some CSR activities won't not demonstrate comes about instantly. Try not to base the achievement or disappointment of your program exclusively on that (Figure 6.1).

Figure 6.1: Corporate social responsibility model focus trends.

Source: Crane et al., (2008)

Don't convey your CSR endeavors in a powerless or sermonizing path, either to your workers or your clients. This will undermine your earnestness and enthusiasm about your CSR activities. Neither your workers nor your clients will get it. In the event that you make a strong CSR program, do it equity by conveying it unhesitatingly and unmistakably.

- Don't give people duty without expert. This remains constant in all parts of business.
- Don't regard CSR as a different organization activity. Consolidate it into all parts of your organization's reasoning.
- Don't quantify without activity. Estimation is a decent initial step. At that point move in the direction of change.
- CSR is costly. Everybody is fixing their belts, and nobody anticipates that little organizations will spend an expansive sum on juvenile CSR endeavors. Roll out little moderate improvements. A few endeavors, such as turning down the indoor regulator, killing the lights and reusing, can really spare organizations cash with a little exertion and insignificant budgetary yield.
- CSR is a trend or promoting contrivance. Studies and talks this year alone demonstrate to us that it's most certainly not a trend.

CHAPTER 7

Current Future Trends and Opportunities for Corporate Social Responsibility on a Global Level?

The old framework breaks separated, as a greater amount of its defects are uncovered. The solitary craving to overwhelm markets is in the long run exceeded by the group will to propel them. Riches and opportunity turn out to be more uniformly circulated all through society. Also, relentless change radiates from the base up. This is the place we are moving right now: into a time of significant market change and drastically unique societal desires (Figure 7.1).

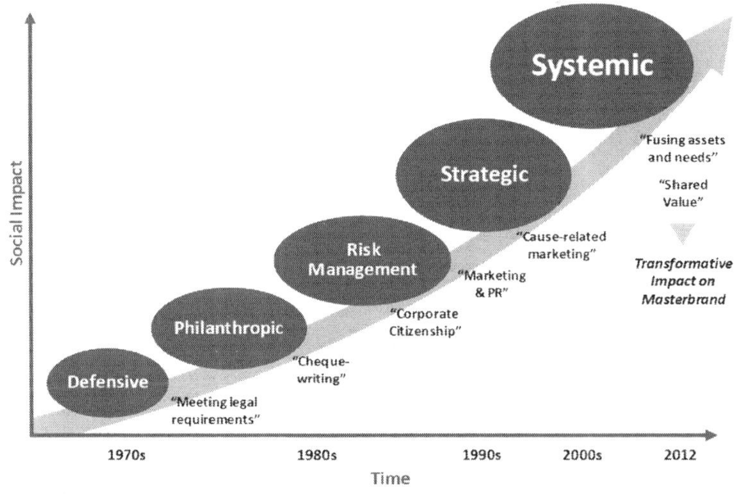

Figure 7.1: Evolving trends in corporate social responsibility in line with societal changes in cultural preferences.

Source: Crane et al., (2008)

While these attributes may appear to be idealistic, there is sufficient proof to recommend that genuine change is in the air. It merits considering exactly how altogether markets may reshape themselves, and why.

As Common Bond CEO David Klein recently told CNBC in an interview: "I think what finance looks like in 10 years is unrecognizable to us today. I think the platforms that are coming to the world are going to completely change the way we think and interact." Very possibly, they already have.

As per the analysis of BSR, during 2017, there are 10 key areas that are reshaping the future of the world. They are discussed in the following subsections.

7.1. GLOBALIZATION UNDER PRESSURE

Rising backing for exchange protectionism and patriotism, reflected in late races and choices in the US and Europe and openly feeling more broadly, difficulties key presumptions about the certainty of globalization.

Moderating worldwide exchange, development of protectionist measures, and quick ascent in nationalistic and xenophobic talk are progressively unmistakable issues all around. The US Presidential race and Brexit, as well as continuous race crusades in Germany and France have all been set apart by patriotism and calls to rollback political, monetary and social advancement. Organizations must decide the scope of boundaries to exchange and develop interdependent streams taking into consideration what these imply for corporate sustainability.

The key signs are:

- The World Trade Organization (WTO) evaluates that worldwide exchange developed by only 1.7% in 2016, its slowest rate since the monetary emergency of 2009. The worldwide body additionally reexamined its figure for 2017, evaluating that exchange will extend at a rate in the vicinity of 1.8% and 3.1%, down from the past estimate of 3.6%. WTO singled out Brexit and "developing against exchange talk" among the components blurring the viewpoint.
- Major global exchange understandings, even those close consummation or as of now set up, are at danger of being deserted. US President Donald Trump marked an official request pulling back the US from the Trans-Pacific Partnership (TPP), an arranged agreements which covers 40% of the world economy.

European pioneers are worried that the Transatlantic Trade and Investment Partnership (TTIP) could be next.

- A new EU-Canada exchange agreements practically fell through after Belgian provincial governments debilitated to hinder the arrangement. Contradictions were in the long run defeat opening the route for the point of interest consent to be agreed upon.
- While a few nations have been striking down hindrances to free development of exchange and capital, this has been outpaced by the rate of new protectionist measures. In 2015 alone, G-20 governments set up 644 unfair exchange measures, and universal loaning has dropped by over 10% since 2014.
- The UK government is because of start arrangements in the not so distant future to leave the EU, after a choice vote to leave the alliance. The transactions will include building up another exchange agreement.
- Analysts caution that the new scene is as of now constraining a few organizations to downsize global operations and is influencing cross-fringe capital streams; however, the long haul effect on maintainable advancement stays to be seen.

7.2. CYBERCRIME EPIDEMIC

The scale and advancement of cybersecurity assaults are developing, putting organizations, governments, and people at expanding hazard. Cyberattacks and information ruptures shook various worldwide organizations and influenced the US Presidential decision in 2016. Machine learning will quicken the number and advancement of assaults in 2017. Internet of things (IoT) malware can possibly open secondary passages into self-governing vehicles and in addition, associated homes and organizations, putting shoppers' close to home information and physical security at hazard. The general population and private cost of cybercrime and the safety efforts intended to forestall it are soaring, and organizations, particularly ICT, energy utilities and car, are under expanding weight to enhance framework security and client information assurance.

The key signs are:

- The cost of cybercrime has risen exponentially (assessed to be $3 trillion in 2015) and could reach $6 trillion yearly by 2021. The related expenses of cybercrime incorporate harm and devastation of information, burglary of protected innovation, and individual and budgetary information, disturbance of business operations, and reputational harm.

- In 2016, Yahoo reported that private information from more than a billion client records was stolen in 2013 by an unspecified programmer, the biggest known information rupture ever. A year ago, programmers kept a great many individuals from getting to prominent destinations and administrations including Reddit and Twitter through a Distributed Denial-of-Service assault.
- Technology, utility and car areas are especially defenseless against cybercrime. In 2016, the US government granted $34 million in financing to ventures went for the assurance of the nation's energy matrix from cyberattacks, and various reports have singled out the transportation segment as being at high hazard.
- By 2020, there are relied upon to be 20.8 billion gadgets associated with the Internet, making private purchaser information progressively powerless and compelling organizations to put resources into more strong information security capacities and enhance approaches.
- As self-ruling vehicles are taken off via automakers, security will be top-of- psyche for organizations. Independent, associated vehicles and IoT are especially helpless against hacking.
- The rising number and multifaceted nature of cybersecurity dangers brought about energy for more noteworthy control. China's new cybersecurity law, and the European Union's General Data Protection direction are probably going to begin a large number of government endeavors to address the ascent of cybercrime.

7.3. CLIMATE LEADERSHIP SHAKE-UP

While the driven Paris Agreement was confirmed in late 2016, the atmosphere administration scene has experienced significant movements since it was arranged, and the political will expected to bolster usage might be progressively hard to assemble and keep up.

While the US assumed a noteworthy part in realizing the Paris Agreement, the Trump Administration has debilitated to haul out of the atmosphere understanding and increment interest in petroleum derivatives.

China is rising as a once-far-fetched champion and driving worldwide interest in sustainable power source, and European nations are proceeding with their support for usage of the Paris objectives.

Also, there is expanded private division movement on atmosphere as found in the number and reality of corporate endeavors on science-based discharges lessening objectives, aspiring sustainable power source targets and enhancing atmosphere hazard divulgence. Local governments and urban areas are likewise standing up and making critical atmosphere responsibilities, recommending authority from subnational on-screen characters and establishments might be what shores up endeavors to convey on the Paris Agreement's guarantee.

The key signs are:

- With arrangements to "unleash America's $50 trillion in undiscovered [fossil fuel] saves," the Trump Presidency may goad expanded US interest in oil and gas. President Trump has likewise guaranteed to rejuvenate the US coal industry; however, request is set for proceeded with decrease in 2017, with China's falling coal utilization anticipated that would additionally diminish interest for the US sends out. Faced with the possibility of the US hauling out of the Paris Agreement, China is venturing up its atmosphere administration endeavors and as of late said that US government activities would not influence its move to turning into a greener economy. China is the world's biggest financial specialist in sustainable power source and as of late reported arrangements to spend in any event $360 billion on sustainable power source by 2020. China likewise plans to dispatch its national carbon exchanging plan in 2017.
- The Canadian government has reported arrangements to have every single Canadian purview present a cost on carbon contamination by 2018. As per Prime Minister Justin Trudeau, areas and domains would have the capacity to put an immediate cost on carbon contamination or settle on a top and- exchange framework. Increasingly, the business group perceives that a steady situation guarantees long haul gainfulness. Several organizations and financial specialists have approached President Trump to precede with US cooperation in the Paris Agreement. More than 200 worldwide organizations so far have focused on science-based discharges decrease targets and keep on taking

activity through aggressive sustainable power source objectives and in addition cooperative coalitions, for example, We Mean Business. The divestment development has additionally been picking up pace. Establishments and people controlling more than $5 trillion in resources have sworn to strip from petroleum products, which speak to 100% development over the most recent 15months.

7.4. SPLASHY DEBUT, UNCERTAIN NEXT STEPS FOR SDGS

Since their dispatch in 2015, the SDGs have pulled in huge corporate, financial specialist and government consideration yet particular activity has been constrained. A little yet prominent number of organizations are openly supporting the United Nations (UN) Sustainable Development Goals (SDGs or the Goals), propelling related activities and focusing on aggressive targets. There is additionally an expansion in the quantity of money related instruments being propelled with the goal of assembling capital lined up with accomplishing the Goals. Be that as it may, boundless responsibility by the private division, also concession to measurements for measuring progress, slacks.

The key signs are:

- Many organizations openly supporting the SDGs still can't seem to set objectives, targets or measurements to track advance, as indicated by a current multi-segment examination of 25 European organizations from PricewaterhouseCoopers.
- According to the Global Opportunity Report 2016, 32% of CEOs studied knew about the SDGs, and 23% were arranging a reaction. In any case, mindfulness additionally down the administration chain was insignificant, with just 5% of mid-level supervisors mindful of arrangements for an organization to follow up on the goals.
- The Business Commission as of late distributed its leader report Better World, Better Business, in which it contends that economic improvement presents novel open doors for organizations and manageable business could open $12 trillion in new market esteem.
- Unilever, SABMiller, Ericsson, and Tata Group are among a

little however developing number of organizations showing solid initiative on the SDGs, adjusting center system with particular Goals and creating quantifiable targets.

- Despite absence of activity, buyer desires of corporate initiative on the SDGs stay high. A current report from Corporate Citizenship found that 81% of millennials trust business has a key part to play in accomplishing the goals.
- The speculator group has additionally been indicating enthusiasm with money-related foundations propelling new activities to bolster the SDGs. UNEP Finance Initiative and Principles for Responsible Investment, speaking to more than 1,500 associations with consolidated $60 trillion in resources under administration, will work with the UN Global Compact to make imaginative money related items that can possibly divert open and private back towards basic foundation and arrangements key to accomplishing the SDGs.
- The World Bank with BNP Paribas will dispatch a program of value file connected bonds, which will enable speculators to add to the financing of supportable improvement ventures and advantage from the execution of organizations making a huge commitment to the SDGs.

7.5. QUALITY OF INFORMATION

The energy of innovation organizations to impact open utilization of data and shape general supposition has achieved new statures.

The ascent of online networking as a noteworthy merchant and some of the time distributor of news working with minimal article or quality control has developing ramifications for government and business. Fake news stories affected the US decision, and with voters making a beeline for the surveys in Germany, France, Norway, the Netherlands, and numerous different nations this year, governments are stressed that created news reports could tip voting results.

The expansion of fake news has constrained innovation organizations to reevaluate the limits of their obligation regarding content shared on their stages, while different organizations are progressively mindful of the genuine ramifications false information may have on corporate brand notoriety.

The key signs are:

- According to the Pew Research Center, over 62% of American adults get their news from an online networking outlet, with 44% of the population getting their news from Facebook.
- New inquire about recommends that Google look calculations can move undecided voter inclinations by as much as20%.
- Norway's driving daily paper has called Facebook CEO Mark Zuckerberg "the world's most intense proofreader;" however, online networking organizations have since a long time ago kept up that as innovation organizations they have constrained obligation regarding content. Notwithstanding, Zuckerberg as of late declared that the organization would find a way to counter fake news, and both Google and Facebook discharged strategies planned to confine income producing promotions from being set on fake-news destinations.
- Pepsi and New Balance experienced direct the effect that fake stories can have on purchaser trust. After created news article misquoting senior officials from both organizations went viral, customers called to blacklist those organizations' items.
- Fake news has likewise been bringing about a tempest outside North America and Europe. It was hailed as a security need for Indonesia, after created stories have exacerbated ethnic and political pressures and set strain on conciliatory relations with China. A fake news story additionally as of late started a Twitter showdown between atomic forces Pakistan and Israel.
- Sweden has approached Facebook to control fake news intentionally or else confront "obligatory measures" by the administration, joining Germany, Finland, and Indonesia in the developing positions of governments that are effectively handling fake news.

7.6. FEEDING THE FUTURE

As the worldwide populace develops and turns out to be more prosperous, issues at the convergence of sustenance, wellbeing, and atmosphere turn out to be progressively basic and confounded. Business for the most part, and nourishment and refreshment organizations specifically, keep on coming under expanding purchaser and administrative weight to address sustenance related wellbeing and atmosphere challenges. As riches in center wage nations rises, infections of prosperity like Type 2 diabetes and heftiness are surging, filled by more noteworthy sugar and meat utilization. Some trust stoutness and its related therapeutic dangers now represent a greater worldwide wellbeing emergency than craving. Creature protein and red meat particularly, are under flame for both their negative wellbeing effects and substantial carbon impression. While meat utilization keeps on expanding in creating nations, there is proof of enthusiasm for the wellbeing and ecological advantages of non-meat proteins in specific markets in Europe and North America.

The key signs are:

- Diabetes rates have expanded by around 400% and corpulence rates have multiplied since 1980, according to the World Health Organization (WHO), due to a limited extent, to the expanded utilization of sugary refreshments in developed nations. Researchers have demonstrated that meat utilization is additionally assuming a basic part in developing worldwide heftiness rates.
- Beverage organizations keep on facing open feedback and developing direction. In 2016, a WHO report asked governments to expense pop with an end goal to address diabetes. Pop duties are relied upon to become effective in South Africa this year, trailed by the United Kingdom and Ireland in 2018.

These nations will join a developing number of governments who have effectively executed sugar-related expenses in the course of the most recent five years, including Mexico, France, Norway, Hungary, and a few US urban communities.

- Companies including PepsiCo and Coca-Cola have reacted to the supported open weight by at the end of the day promising to cut the sugar and calorie substance of their items.
- The animal industry is a noteworthy supporter of our worldwide carbon impression, representing around 18% of aggregate ozone-harming substance emanations. Accordingly, nations, for example, Denmark are thinking about an assessment on red meat to battle environmental change.
- The Chinese government arrangements to cut meat utilization by half among its 1.4 billion subjects. The measure is intended to enhance general wellbeing results yet could likewise give an essential lessening in ozone-harming substance emanations.

- A developing number of organizations, including one of the world's biggest meat makers, Tyson Foods, are reacting to customer interest for non-meat proteins in Europe and North America. This year will see promote development in plant-based nourishments as clients organize wellbeing and health.

7.7. FINANCIAL DISCLOSURE REQUIREMENTS EMERGING

A scope of new approaches and proposals from controllers and standard-setting bodies both guide and drive more prominent organization exposure on sustainability execution.

Speculators of numerous types are progressively focusing on supportability, and a developing number of monetary organizations have units committed to practical contributing. While getting too convenient and vigorous information on ecological and social execution remains an obstruction to more quick development here, there is perpetually and better revelation direction accessible. Various controllers and standard-setting bodies have discharged or are taking a shot at new divulgence rules, and a developing number of organizations are confronting weight from financial specialists to be more straightforward about manageability dangers and execution. The key signs are:

- According to the US SIF Foundation, in 2016 more than one out of each five dollars under expert administration in the US – $8.72 at least trillion – was contributed by reasonable systems, a number which has reliably developed amid the most recent decade, and is up 33% contrasted with 2014. Europe has additionally observed take-up in feasible contributing of various types.
- The Financial Stability Board's Task Force on Climate-related Financial Disclosures has discharged suggestions for organizations on environmental change exposure to be made as a component of their standard monetary filings. This will make it less demanding for financial specialists to join atmosphere related dangers in their choices.
- Momentum has likewise been working for stock trades to survey their prerequisites for exposure of ecological and social issues. The lion's share of remarks submitted in a current round of open discussion about exposure rules by the US Securities Exchange

Commission highlighted the requirement for more revelation on ecological, social and administration (ESG) issues.

- FTSE Russell, a worldwide pioneer in value files, arrangements to discharge ESG announcing direction for backers of both the value and security markets.
- Companies are feeling more weight from shareholders to reveal information identified with manageability dangers. Resolutions submitted a year ago to Chevron, and Exxon Mobil called for stress tests to decide the hazard that endeavors to check environmental change stance to their organizations. The way that the resolutions accomplished 41% and 38% votes separately exhibits the expanding noteworthiness with which institutional financial specialists see environmental change.

7.8. ACCESS, ANTIBIOTICS, AND PANDEMICS

Worldwide wellbeing establishments from multilaterals to national governments and additionally organizations keep on finding access to drugs, anti-microbial resistance and pandemics the most intense difficulties to enhance human wellbeing. Ebola and Zika, as most different pandemics that went before them, shocked the world, underscoring the absence of readiness of the worldwide group to bargain adequately with pandemics. As multilateral associations and governments keep on working on enhancing availability for such emergencies, organizations confront expanding weight to put resources into arrangements. Such proactive arranging, alongside enhanced access to medications and battling anti-infection resistance will stay focal wellbeing topics impacting the corporate supportability plan of pharmaceutical organizations specifically.

The key signs are:

- The late pestilence of Ebola guaranteed more than 11,000 lives, and the flare-up of Zika has been connected to a great many children conceived with immature brains. These pandemics got the worldwide group off guard the race to create antibodies is currently at full speed. Merck and Co, with the help of a general wellbeing office in Canada, built up an exploratory antibody for Ebola that in a current trial was appeared to give 100% insurance, and GlaxoSmithKline, Sanofi, and others are effectively taking a shot at a Zika immunization.

- Access to drugs remains a noteworthy worldwide wellbeing worry with more than 2 billion individuals as yet inadequate with regards to access to medication. As indicated by the latest Access to Medicines Index, GlaxoSmithKline, Johnson and Johnson and Novartis are driving worldwide endeavors in this field.
- In an advancement seen as a noteworthy stride to enhance worldwide access to meds, the UN Human Rights Council received a milestone determination, supported by a gathering of creating nations that approaches states to apply the standard of delinking restorative R&D from the costs of solutions, diagnostics, and antibodies.
- The abuse of anti-toxins and the spread of safe super-bugs have turned into a suggestion to take action for business to take a shot at arrangements. Researchers have cautioned that the world is on the cusp of a "post-anti-toxin period," otherwise called the "anti-toxin end of the world," when microscopic organisms will end up noticeably impervious to known medications. More than 700,000 individuals are assessed to pass on every year from medication safe contaminations. This number is anticipated to achieve 10 million by 2050.
- Reacting to an invitation to take action by the UN, 13 driving pharmaceutical organizations as of late exhibited another industry guide that lays out four key duties by the private area to be actualized by 2020 to battle antimicrobial resistance.

7.9. EVER MORE CONSUMPTION

Prodded by expanded success and populace development, rising utilization rates in China and other creating markets produce monetary development and supportability challenges in equivalent measure. Crosswise over developing economies, expanding riches and financial development are being joined by developing utilization and asset utilize. Notwithstanding late stoppage in GDP development, China's utilization rates keep on rising at a quick rate. Most rising economies' development is outpacing the limit of their waste administration frameworks, quickening impacts on human wellbeing, land, and marine biological communities.

Organizations working in developing markets have a one of a kind chance to band together with nearby groups on commonly useful answers for handle the negative social, financial and ecological effects of fast development.

The key signs are:

- E-business has encountered quick development in China. In 2016, web-based shopping made up 15% of aggregate private utilization, a sharp increment since 2010, when it was just 3%.
- Chinese car and gadgets areas have likewise observed fast extension. Auto deals are developing at the yearly rate of over 10%, and telecoms mammoth Huawei gauges that incomes from its shopper gadgets division ascended by around half in 2016.
- On China's Singles Day in 2016, Alibaba's yearly web-based shopping occasion accomplished a record $17.8 billion in deals in only 24 hours – beating the records of earlier years and surpassing the consolidated offers of comparative US occasions, Black Friday and Cyber Monday, and raising worries of the ecological effects of bundling waste.
- Lifecycle effects of items will be an undeniably pivotal issue for organizations to handle, as the Chinese government increase control with the Extended Producer Responsibility arrange, anticipated that would happen in 2025. In Latin America, Argentina and Mexico are handling developing waste difficulties with proposed controls focusing on bundling, e-waste, and polystyrene.
- A late review by the Ocean Conservancy found that the dominant part of plastic enters the sea flanking five quickly developing economies – China, Indonesia, the Philippines, Thailand, and Vietnam. These nations have encountered detonating interest for buyer items yet squander administration foundation has neglected to keep up.
- Waste is likewise a developing test for some African nations. In Ghana, general society plastic waste issue has been significant to the point that the legislature reported a halfway prohibition on light plastics in mid-2015. Nonetheless, reaction from business constrained the legislature to surrender the arrangement and rather propose another law that requires the utilization of biodegradable plastics.
- Multinational organizations have beforehand centered around "social permit to work" in developing markets, however, are currently step by step moving towards a "permit to develop" methodology, which incorporates measures to address

neighborhood ecological and social difficulties. Unilever is one such organization that has moved to a long haul development system in developing markets, where it sees lessening ecological effects, for example, water utilization, and tending to social difficulties, for example, cleanliness and sanitation, as key to its proceeded with development.

7.10. A JOBLESS FUTURE DRAWS CLOSE

Progresses in counterfeit consciousness, mechanical autonomy and different innovations are bringing about significant moves in the work advertise, requesting governments and organizations to think about inquiries concerning innovation's effect on employment and the economy. As advances, for example, mechanical technology, manmade brainpower, and sensors progress and move toward becoming standard, the nature of work is quickly evolving. Worldwide goliaths, for example, Amazon and Uber, which at present give employments to a huge number of individuals, are progressively coordinating computerization in their operations, supplanting people with machines. A few nations are because of dispatch national investigations of fundamental pay as a way to give general wage as economies change through computerization. While evaluations of how much the creation of robots will influence future business contrast, specialists concur that the future workforce is probably going to recoil and the way of occupations that remain will change.

The key signs are:

- Robots will assume control 6% of US employments by 2021, beginning with client benefit agents, and taxi and truck drivers, as indicated by a report by statistical surveying firm Forrester. A report by Oxford University goes significantly, additionally recommending that as much as 47% of all employments will be computerized by 2035.
- More than 137 million specialists in Southeast Asia could lose their business to computerization in the following 20 years, as per a 2016 report led by the International Labor Organization.
- Finland has turned into the main nation to dispatch a national pilot of fundamental wage, which will give an unequivocal month to month whole to 2,000 unemployed Finns for a long time to supplant existing social advantages. Defenders of an essential salary contend that it will guarantee better riches redistribution

in the economies quickly changing because of occupation robotization, ensure individuals against employment frailty and cut formality related with social advantages. Italy, the Netherlands, and Canada are among the nations that are additionally propelling comparative pilots.

- Uber has begun guiding driverless autos in Pittsburg and San Francisco and arrangements to inevitably supplant the lion's share of its drivers with computerized autos. The organization has likewise as of late obtained a self- driving truck start-up, Otto, and is hoping to dispatch a self-driving whole deal trucking administration.
- If Amazon's without registration shop means that things to come, the retail division ought to prop for difficult task misfortunes. The worldwide retail goliath is guiding a genuine shop in Seattle, which enables clients to stroll in, take what they need, and exit. Their records are charged later, naturally.
- Farming is another segment encountering significant movements. Rambles, computerized reasoning, and sensors are progressively connected in cultivating to plant seedlings, furrow the fields, sort, and bundle deliver and do numerous different employments that supplant human work.

CHAPTER 8

Current Future Trends and Opportunities for Corporate Social Responsibility on a Management Level?

More on a supportability administration level as indicated by the corporate world, there are 10 rising patterns:

The accompanying rundown of 10 patterns for 2015 distills Sustainability's reasoning about these and numerous different issues over the previous year and gauges the issues that will shape the manageable advancement motivation in 2015.

8.1. SCALES TIP ON WORLDWIDE ENVIRONMENTAL CHANGE ACTIVITY

With rising community activism, surging quantities of corporate duties and more conclusive activity by neighborhood and national governments, worldwide environmental change strategy is hinting at a new development trend. Changes within government environmental policy is crucial as well as implementation of worldwide pacts and frameworks where both developed and developing countries have their set responsibilities and accountabilities.

What's next?: There is certified – albeit still mindful – idealism that the global group at long last may concede to a significant and enduring system for de-carbonization in 2015.

- In December, 200 nations marked the Lima Accord, which without precedent for history confers all countries to cutting outflows. In mid-2015, each nation will submit a nitty gritty household approach arranges that will shape the premise of the bargain to be marked in Paris one year from now.
- The U.S. what's more, China struck the main significant

arrangement between an industrialized and a creating nation on environmental change. The U.S. vowed to slice discharges 26–28% beneath 2005 levels by 2025, while China consented to top outflows around2030.

- More than 25 nations have vowed $10 billion to the Green Climate Fund, set up to help creating countries adjust to environmental change.

8.2. WATER MAKES WAVES

From dry spells in California and Ohio to proceeding with water deficiencies in India and Brazil's Sao Paolo, the effect of water shortage on populaces, economies, and organizations have been felt around the world. Worldwide water stress is quick turning into a noteworthy monetary, political and social issue, and an inventory network hazard to which organizations are reacting with a scope of activities and developments.

What's next?: As weight to address the test works in both the general population and private areas, 2015 will see considerably more purposeful activity to comprehend and follow up on water-related issues with an accentuation on cooperation and framework level arrangements.

- The World Economic Forum distinguished "water emergencies" as one of the main 10 issues of most noteworthy worry to the worldwide economy in 2014, the third most noteworthy hazard positioned in general.
- A developing number of organizations perceive water shortage hazards as material and are including them as a monetary worry in yearly reports. For instance, Dr. Pepper Snapple Group noticed that water supply difficulties could put $2.5 billion of its future deals at hazard.
- Investments in water framework improvement including watershed installments, water chance administration instruments, customer engagement programs, and natural effect bonds are developing as inventive and promising approaches to lighten, and maybe advantage from, the water shortage challenge.

8.3. OFFICEHOLDERS CONFRONT A DEVELOPING NUMBER OF TENABLE DANGERS

New advancements, moving buyer inclinations and a developing development of social business people and different trendsetters are upsetting conventional plans of action and definitely adjusting existing markets. In 2014, organizations in various industry parts – including car, electric utilities, saving money, retail and nourishment– started grappling with progressively reasonable dangers to their organizations.

What's next?: While not the greater part of this disturbance has sustainability as its objective, another era of items and administrations – and entire new methodologies, for example, the sharing economy – are finding expanded validity and effect and making the idea of a more feasible economy more substantial.

- Transportation is experiencing radical change as organizations, for example, Uber, BlaBlaCar, and Lyft, notwithstanding the blast of bicycle share programs, have customers reconsidering how they travel. Disturbance from inside is likewise quickening as electric autos turn out to be less puzzling and more affordable and further open customer psyches to more economical advancement.
- Utility organizations and apparatus producers must fight with problematic innovations presented by newcomers, for example, Apple and Google, which are trying different things with approaches to make home energy utilize more proficient.
- Crowdsourcing of back led by such pioneers as GoFundMe, Kickstarter, and Indiegogo is testing the saving money industry's current model and offering better approaches to give funding to imaginative arrangements.

8.4. THE CLOCK IS TICKING FOR ENERGY ORGANIZATIONS

While vulnerability keeps on commanding the future standpoint of the energy segment, record picks up by renewables, the ascent of the petroleum derivative divestment development and developing force for a worldwide environmental change accord are setting weight on conventional energy organizations to exhibit that they're not kidding about a worldwide temperature alteration.

Coal, oil, and gas organizations recognize environmental change- related dangers, yet few have made adequate move to flag their acknowledgment of and responsibility regarding the need to move.

- *What's next?:* Falling oil costs are probably going to briefly moderate the quick climb of renewables, yet that won't mitigate oil, gas and coal makers from the need to reexamine their parts in the rising low-carbon economy.
- The International Energy Agency (IEA) predicts that sun oriented will turn into the overwhelming wellspring of power by 2050. Sustainable power source made record picks up in 2014 with drops in innovation expenses and fast extension of limit and speculations.
- The Energy Company EO, stated that it would part its operations to concentrate on clean energy raises the possibility of progress for Germany's Energiewende program that intends to altogether wipe out non-renewable energy sources.
- The petroleum product divestment development saw picks up in 2014, with expanded municipal interest and more than $50 billion stripped by more than 180 foundations and near 700 people. The thought of "stranded resources" is picking up acknowledgment as a hazard considers influencing standard speculation choices.

8.5. RETHINKING A 'FUTURE FIT' WORKING ENVIRONMENT

Moving worldwide socioeconomics are convincing organizations to reconfigure work rehearses and suits a multigenerational and progressively differing workforce. While living compensation questions keep on dominating the discussion about reasonable business rehearses, organizations are perceiving that wages and contracts are just a piece of the Astound.

What's next?: Workplaces that are fit for the future will execute the more extensive drivers of prosperity, for example, adaptable working hours and calendars, terms of agreements that are advantageous for both managers and representatives, bigger interests in preparing and instruction and flexibility to representatives working until a later age.

- Forward-looking organizations perceive the aptitudes and diligence of more established laborers and are creating inventive approaches to oblige their necessities, including offering half-retirement contracts, organizing senior specialist abilities in enlisting, making new positions and changing ergonomic settings.

- To address the issues of more youthful laborers, Starbucks declared that it would give subsidizing to workers to get school training. Accordingly, it likewise reported overhauling its booking programming to empower directors to make work move plans more family-accommodating.
- Malaysia is putting forth organizations charge motivators on the off chance that they build up nurseries and take into account adaptable working hours along these lines empowering moms to take an interest in the workforce.

8.6. CORPORATE DIFFERING QUALITIES PAST SEXUAL ORIENTATION

Despite the fact that sex keeps on driving the workforce differences discussion, partners and organizations are turning their concentration to extra measurements of incorporation, for example, race, ethnicity and sexual introduction. Innovation firms made a major jump in 2014, revealing representative information reflecting sexual orientation, racial and ethnic underrepresentation and recognizing the need to change.

What's next?: In 2015, organizations will be relied upon to show their sense of duty regarding differing qualities through yearning targets and objective setting; however, in the event that Apple's CEO turning out as gay to convey perceivability to the non-segregation issue is any sign, the corporate division still has much to do make differences some portion of its authoritative DNA.

- Silicon Valley mammoths Yahoo, Google, Facebook, Twitter, and Amazon stood out as truly newsworthy in 2014 after at last uncovering their assorted qualities figures. Tech organizations are under expanding weight from partners, including social liberties lobbyist Jesse Jackson, to handle the transcendently "white male" synthesis of their workforces.
- Apple CEO Tim Cook's turning out freely highlighted the degree to which LGBT segregation remains an uncertain and once in a while talked about issue in the corporate world.
- Next year will see more organizations declaring assorted qualities focuses on that go past sexual orientation. For instance, KPMG's

objectives will be set crosswise over sex, race, inability and sexual introduction.

8.7. WEIGHT ASCENDS ON ORGANIZATIONS TO KILL SUBJUGATION

The quantity of worldwide human rights infringement have been ascending as of late, with a substantial share of encroachments ascribed to specialists' rights infringement, arrive snatches and illicit store network hones in developing markets. As partners call attention to the constraints of store network reviews, organizations have been forced to raise responsibility benchmarks and execute new measures.

What's next?: In 2015, the U.K. is set to pass the Modern Slavery Bill, one of the main laws of its kind on the planet, yet it will take numerous more years of deliberate support by partners to see a move from intentional to mandatory administrative structures on subjection and other human rights issues?

- Human rights infringements have ascended by 70% over the most recent six years, as indicated by Human Rights Watch and Maplecroft. Government constraint and clashes have been key elements representing the expansion yet numerous infringements are additionally ascribed to corporate activities.
- The Modern Slavery Bill that was discussed in the U.K. Parliament will oblige organizations to report endeavors towards killing present day subjection from their supply chains.
- Pope Francis has called for more activity to battle human trafficking and subjection by engaging customers to not purchase products made by misused specialists. The International Labor Organization assesses that constrained work produces $150 billion in illicit benefits every year.

8.8. MORALS DEFICIENCY FREQUENTS BANKS AND ORGANIZATIONS

Inquiries regarding the condition of morals keep on plaguing vast multinationals and banks taking after a year set apart by pay off embarrassments, punishments on the fund area and reports highlighting the absence of straightforwardness crosswise over businesses. Different organizations have battled allegations of abstaining from paying what's coming to them of duties by moving assessment houses abroad.

What's next?: While European and U.S. governments have made strides in 2014 to brace down on expense reversal, they will present considerably more grounded measures in 2015. All things considered, it will take numerous years for controllers, partners and organizations to move the needle on the issue.

- In what is turning into an undeniably typical example, Barclays, Bank of America, J.P. Morgan and HSBC were among numerous money related organizations blamed for twisting the standards to make benefit and that confronted critical administrative fines and different punishments subsequently.
- OECD nations concurred on a progression of measures to decrease "assess shirking by multinational organizations" and the U.S. government provided new principles too. Burger King, Pfizer, and Walgreens needed to fight off allegations of endeavoring assessment reversal.
- GlaxoSmithKline and Tesco endured hits to their notorieties after, separately, being discovered blameworthy of remuneration by Chinese courts and exaggerating benefits.

8.9. ORGANIZATIONS SET FOR A BIGGER PART IN GLOBAL ADVANCEMENT

As the isolating lines amongst industrialized and developing markets obscure, multinational companies logically have been advancing financial development in the creating scene and adding to answers for complex worldwide issues, for example, neediness. The part of organizations is set to additionally quicken as U.K. organizations and the E.U. reinforce their emphasis on the cooperation of business in universal improvement.

What's next?: The United Nations (UN) will finish 17 Sustainable Development Goals (SDGs) and their execution will be an essential trial of the private division's hunger and inclination for bigger contribution in worldwide social, political and financial undertakings.

- The UN will finish an arrangement of 17 SDGs in 2015. The proposed objectives cover wide topics: finishing neediness and hunger and enhancing wellbeing, instruction, and sexual orientation correspondence. They additionally incorporate particular focuses for lessening imbalance, making urban communities safe and tending to environmental change.

- The U.N. Worldwide Compact is looking into ways that organizations and monetary foundations can bolster needs of the post-2015 plan. Together with the Global Reporting Initiative (GRI), the U.N. Worldwide Compact has made a universal system of experts and is working with groundbreaking organizations to execute a guide that backings organizations in evaluating their effects, adjusting their techniques to the SDGs and setting organization objectives.

8.10. THE GUARANTEE OF BUYER ENGAGEMENT

Corporate manageability pioneers have perceived that accomplishing durable change requires coming to past companywide activities. As enterprises search for creative approaches to draw in buyers in positive conduct change, millennials are a key statistic, with the quantity of youth engagement activities on supportability and natural issues developing consistently.

What's next?: With the expanding utilization of online networking stages, we will see more development in this circle in 2015 – despite the fact that not all examination underpins the claim that more profound engagement prompts more feasible utilization designs. Organizations will search for more grounded confirmation to go down promoting activities as they likewise grow endeavors gone for other age gatherings.

Twenty-nine driving partnerships and charities, including Coca-Cola, IKEA, and PepsiCo, have united to dispatch "On the whole," an advanced stage expected to draw in millennials in economical ways of life.

Several organizations have distributed research about the achievement of promoting activities particularly focused at millennials. In any case, discoveries are blended with some demonstrating millennials as key drivers in supportable utilization while others demonstrate no critical varieties crosswise over age gatherings.

The quantity of applications that empower customers to live more maintainable ways of life and settle on moral purchasing choices has seen colossal development, from data on item accreditations to the accessibility of sharing economy administrations.

CHAPTER 9

A Historical Perspective

CSR is synonymous with terms, for example, corporate obligation, corporate citizenship, sustainable mindful business (SRB) and corporate social execution. The idea of CSR was customarily restricted to "generous giving." However, finished the 1950s, enterprises completed operations with a determined concentrate without anyone else interests. Be that as it may, by the 1960s, the situation changed when organizations were made aware of their duties toward society past their legitimate commitments. Issues, for example, contamination and overpopulation were conveyed to the fore.

Incorporate "profit, give altruism, secure the earth and shield one's items." However, the consequent decades saw a progression of corporate embarrassments. Thusly, specialists needed to move accentuation to advancing decent variety, securing specialists, counteracting kid work, cultivating general wellbeing, guaranteeing human rights, annihilating neediness, contradicting degenerate government administrations, giving innovation, policing supply chains, drawing in partners, estimating and revealing, and enhancing procedures and approaches by and large.

The evolution of CSR practices in India can be traced through the timeline given in Figure 9.1.

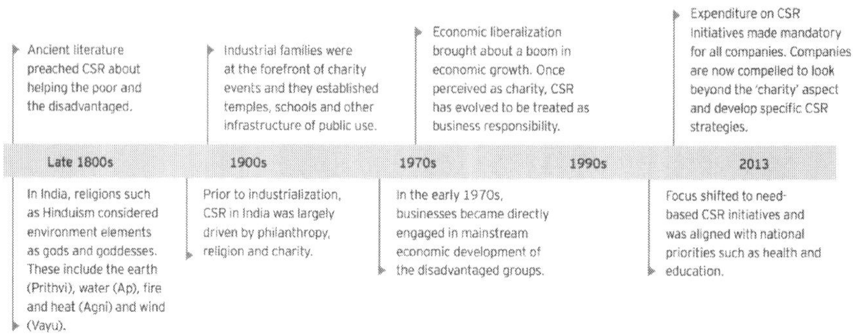

Figure 9.1: Evolution CSR practices in India.

Source: Ernst & Young (2017)

CHAPTER 10

Key Challenges for Implementing Sustainable Business Practices

Without a doubt, there is a big interest and investment in sustainability and socially conscious companies. However, before they start their excursion, it is likewise imperative for businesses to comprehend the genuine significance of sustainability.

A sustainable association is one that at the same time conveys on social, monetary and ecological components to guarantee long haul improvement for the general public and for businesses. Basically, a practical business can work along all mainstays of the triple bottom line approach.

10.1. INDIVIDUALS INTERRELATIONS

They comprehend that individuals bring along certain requirements, circumstances or beliefs. Desires from various partners, for example, employees, investors, clients, sellers, government and groups need to be guaranteeing so that they:

- Create a positive effect on nature.
- Create a positive effect on society and neighborhood groups that they work in. Even though sustainability practices have been proven to be very useful and necessary, the main challenge within organizations and the reason why sustainability strategies are not implement is: Lack of management expertise: Most senior executives see sustainability as "decent to do" versus "need to do."

Often, the authority does not see a connection between endeavors toward social and financial improvement and this is reflected on CSR practices:

- Given the long-term effectiveness of sustainability, businesses feel that the economic investment required is not payed-off.
- Limited comprehension and association of sustainability upon the different organization levels: Companies oftenly neglect to update business practices at all levels of the organization. There is no appropriate relation between the strategy with the workers or coordination among different partners.

This can bring about weakening of the targets of sustainability exercises. On the off chance that we nearly take a gander at above obstacles, the HR capacity can assume a reactant part in promoting change and encouraging the usage of sustainable business practices.

CHAPTER 11

Sustainability and Business

In the a long time since Ceres started performing research and analysis of global CSR practices, they have firmly held to the rule that tending to the world's most noteworthy ecological and social difficulties is a basic for the corporate and money related groups and that inability to address these difficulties risks our capacity to make flourishing in the long haul. Ceres has made key building hinders for meshing ecological and social difficulties into center business practices to accomplish sustainability.

They started with the Ceres Principles, a 10-point code of corporate natural principals drafted in light of the Exxon Valdez oil leaks. At that point, they propelled the Global Reporting Initiative (GRI), now the universal standard for sustainability reporting, utilized by more than 1,100 organizations around the world. They likewise presented the idea of "environmental risk" – now profoundly incorporated into the corporate and financial specialist vocabulary.

These activities have advanced us all, yet more should be done to deliver the outcomes required. Incremental advance in handling worldwide environmental change and other sustainability dangers isn't sufficient. We require quickened execution upgrades from organizations that mirror the genuine logical and financial effects of unchecked carbon contamination, decreasing water shortage and billions of individuals working under poor conditions. Business is sharp at taking care of issues, and a significant number of the greatest worldwide difficulties we confront are social and environmental. Subsequently, it is business that must lead the route by transforming these difficulties into opportunities. This implies completely incorporating sustainability contemplations into administration, execution, responsibility, R&D and general business technique. Following outcomes, breaking down information and executing activities to build productivity and intensity are foundations for progress.

The main issue: sustainability must be the establishment of the 21st-century enterprise. A year ago, Ceres disclosed a striking vision, Ceres 20·20 for accomplishing a sustainable worldwide economy by 2020. It mirrors the solemn reality that our accomplishments to date are not enough – that organizations and financial specialists must do fundamentally more to genuinely adjust their plans of action and venture systems with the intense arrangements expected to guarantee a prosperous and sustainable future.

Ceres 20·20 lays out four key columns to accomplish worldwide sustainability: guaranteeing legitimate bookkeeping to mirror contamination's actual costs; setting new models and desires for business administration; quickening green development; and changing the principles of the amusement so sustainable organizations can contend on a level playing field.

11.1. THE 21ST CENTURY CORPORATION

The Ceres Roadmap for Sustainability concentrates on the second pillar – setting new measures and desires for business authority. It is a manual for organizations on their adventure to reach sustainability – from the meeting room to the common room – and all through the company network.

The report has 20 key desires identified with administration, partner engagement, divulgence, and execution. Inside administration, they support board decent variety and sustainability aptitude, and official remuneration bundles that line up with sustainability execution.

In the employee competences area, we empower revealing of information and objectives for key natural and social difficulties. Under partner engagement, they energize "opening the entryways" to financial specialists, non-legislative businesses (NGOs), and different gatherings concentrated on an organization's sustainability techniques. The guide focuses on execution rather than planning. On measure to getting things done.

Organizations must create substantial outcomes that put us on a really feasible path. For environmental change, that implies a 50% change in energy effectiveness and a 25% decrease in carbon emissions by 2020. It implies wiping out perilous waste, having shut circle frameworks set up, diminishing the impression of providers and expanding human rights models by 2020. Execution will be a definitive measure for assessing an organization's advance towards accomplishing sustainability.

The best performing organizations of the 21st century will be those that perceive this advancing new request, and contribute and act now. These organizations will be best situated to flourish in the coming low-carbon, asset compelled worldwide economy of the 21st century. The race for sustainability is on.

"We simply cannot afford to wait any longer. Our planet's natural resources are depleting at unprecedented rates. We are already seeing real leadership from some of the largest companies around the world. The time has come for all in the business community to stand up, take meaningful action, and become part of the solution."

—Frances Beinecke, President, Natural Resources Defense Council

"Integrating sustainability is not just a good opportunity for business. It is essential for success in a world of constrained resources. Right now every business has a choice to make. We choose to move fast, using sustainability as a force for innovation. We choose to embrace transparency, collaboration, and advocacy as tools to unlock opportunity and enable us to thrive in a clean and green economy."

—Mark Parker, CEO and President, Nike

"We expect our portfolio companies to do what is necessary to position themselves for a sustainable economy. Environmental, social and governance issues are core to business performance. We are looking for companies that are managing these risks, developing business opportunities, and disclosing their results."

—Anne Stausboll, CEO, California Public Employees' Retirement System

"Our economic future depends upon establishing a low-carbon energy system. We need to dramatically increase our investments in energy efficiency, renewable energy, smart grid technologies, and other innovations. These investments will ensure that our future is not only sustainable, but prosperous."

—Peter Darbee, Chairman of the Board, CEO and President, PG&E Corporation

The race is on to build up a low-carbon feasible worldwide economy and the opposition is furious. Trillions of dollars and the prosperity of billions of individuals are in question, and in addition the wellbeing of the planet that supports us.

Organizations and the capital markets are in the middle of everyone's attention, and achievement will rely upon their capacity to put the Earth and its best interest at the center of corporate methodologies. It is an urgent moment with colossal opportunities – a moment that requests perfection in corporate authority, vision and advancement.

Front line advances and the data transformation have changed the way we live, quickened the spread of data and catalyzed financial development.

Open markets have made immense business openings while lifting expectations for everyday comforts for some – however not all – over the globe. At no other time have we encountered such hazardous change and phenomenal development inside the traverse of only one lifetime. In any case, this development has included some major disadvantages. Our atmosphere is warming because of human conduct. Here and now thinking has left us with a worldwide subsidence.

Regularly expanding pollution and water deficiencies are undermining governments, invigorating clash and fueling worldwide neediness. Total population, officially stressing by restricted access to assets, will top 9 billion out of 2050. Business advancement and the energy of capital markets are expected to address these difficulties.

Organizations on the main edge are reacting with another business worldview that makes sustainability an execution a linchpin for future development. They perceive that "the same old thing" is not any more adequate and that specially appointed sustainability activities are never again adequate. Achievement requires setting sustainability at the epicenter of plans of action. Natural, social and administration issues must be flawlessly coordinated into vital arranging and basic leadership.

Organizations must mirror an understanding that they are reliant upon products and ventures gave by nature, and that nature's points of confinement and limited assets must be completely esteemed and overseen for long-term development and success. Organizations, fundamentally, must lead the drive to a sustainable worldwide future and they have to begin today, not tomorrow. Deferral isn't an alternative. Tremendous opportunities emerge amid transformative circumstances. The $6 trillion energy industry – six times bigger than the Internet economy – must be retooled to limit energy consumption and to accomplish a considerably reduction in carbon footprint.

Clean energy production innovations economies and organizations that put themselves out in front will profit the most. Organizations with items and administrations sensitive to the new economy will develop as champs. Addressing the requirements of billions of individuals in developing markets speaks to a considerable duty and test for business.

Huge numbers of the world's 4 billion poorest – are moving past subsistence to dynamic interest in capitalist economies, yet their crucial needs – for clean water, better nourishment, energy, human rights, and versatility – are not being met. This hole presents enormous opportunities – evaluated at $5 trillion a year – for organizations that are situated to develop and convey ease, economical innovations and services.

Investors are becoming progressively mindful of the dangers that environmental change, water shortage, work environment conditions and other natural and social issues present to organizations' primary concerns. They are advising organizations they put resources into to react with forceful methodologies that change chance into circumstance.

Effectively, many organizations are reconsidering their plans of action to address these quick advancing dangers and openings. The greater part of worldwide officials in a 2008 study by the Economist Intelligence Unit thought about corporate social duty (an equivalent word for sustainability) a high or high need – up from 34.1% out of 2005.

A developing number of organizations are attesting initiative on sustainability execution to separate themselves from their companions.

Ceres' vision and desires for corporate accepted procedures in the coming years – objectives that must come to speak to the standard, not the special case. The sustainability journey has just started for some organizations and for the individuals who are simply beginning; they offer a practical and clear guide to quicken their endeavors. It is planned to challenge, and additionally educate and help, the individuals who expect to incorporate sustainability into their business. It investigates the method of reasoning and key contemplations associated with making the move to sustainability, points of interest procedures and instruments being utilized by a few organizations, and gives recommendations to the up and coming age of best practice. The paper lays out four wide zones of movement that organizations should concentrate on and accomplish by 2020. Those zones incorporate administration, partner engagement, exposure, and execution. While the consideration given to every segment relies on the specific business and industry, we trust this guide will demonstrate precious to all organizations. The main segment plots center administration building squares – administration structures, objective setting, and key basic leadership forms – that are expected to coordinate sustainability.

Organizations can utilize proactive partner engagement, featured in the following segment, to survey the relative significance of particular objectives and the adequacy of systems. Revealing basic data to partners, as indicated by the desires set out in the third area, will help demonstrate that an organization's sense of duty regarding sustainability is genuine and its execution solid.

Finally, the guide subtle elements key execution regions for estimating how organizations are advancing on sustainability. It incorporates requesting execution desires with regards to the scale and direness of the sustainability challenges before us.

Presently, like never before, organizations must start adjusting their execution objectives against national and global execution principles that are grounded in science. On account of environmental change, for instance, organizations must lessen their ozone-depleting substance emanations by no less than 25% beneath 2005 levels by 2020 keeping in mind the end goal to meet decreases that driving researchers concur are important to anticipate cataclysmic warming. Just as sound business choices must be founded on science, it is likewise critical for organizations to react to societal desires.

It has turned out to be certain that it isn't adequate anyplace on the planet to deliver products in perilous or exploitative conditions. These are genuine business dangers for worldwide organizations. In the coming years, the techniques that organizations seek after will decide their investor esteem, as well as the fate of our species and our planet. It is immediately an overwhelming test and a tremendous open door.

CHAPTER 12

Key Drivers of Sustainability

12.1. RIVALRY FOR ASSETS

The total population is anticipated to increment to more than 9 billion individuals by 2050. Rising expectations for everyday comforts will bring about both extended markets for merchandise and ventures and extraordinary requests on the planet's normal assets. A significant number of the assets once considered renewable – like woodlands and water – have turned out to be limited when we consider that human requests are developing more rapidly than the capacity of normal procedures to recharge them.

While fatigue of wares can be checked and estimated, the effect of consumption on biological systems is harder to measure and frequently difficult to cure. With asset, exhaustion comes danger of contention as individual's battle to meet their fundamental needs. Take water – population development, financial advancement, and environmental change are stressing access to new water all inclusive.

By 2025, 66% of the total population will live in water-focused on nations, posturing critical dangers to the monetary and social solidness of whole areas and to the corporate operations in those districts.

12.2. ENVIRONMENTAL CHANGE

Our flow petroleum product based economy has prompted a developing convergence of ozone-harming substances in the climate that is driving more extraordinary climate occasions, more extreme and incessant cycles of dry season and surge, and rising ocean levels. These marvels are being met with new approaches and directions, including those intended to utmost and put a cost on carbon outflows. Organizations need to get ready for an arrangement domain progressively threatening toward carbon emanations and for the expenses of adjustment to environmental change. Countless and financial specialists have met up to approach governments at the national and worldwide level to execute thorough atmosphere strategy.

These groups include Business for Innovative Climate and Energy Policy (BICEP), US CAP, The Prince of Wales Corporate Leaders Group on Climate Change, the Investor Network on Climate Risk (INCR) and the Institutional Investors Group on Climate Change (IIGCC), among others. These businesses recognize the opportunity to profit from technologies that reduce emissions and create solutions to global warming.

12.3. ECONOMIC GLOBALIZATION

The integration of national economies into the global economy brings opportunities for business, but often with significant risks. More and more companies operate in or source from multiple countries with wide disparities in enforced environmental and social standards. Whatever the local enforced standard, many stakeholder groups demand, at a minimum, that companies meet international expectations.

12.4. CONNECTIVITY AND COMMUNICATIONS

Advances in digital communication over the last two decades have reduced not only the time it takes to build a reputation, but also the time it takes to destroy one.

Communication is increasingly disaggregated across multiple social networks. Facebook has over 65 million users, and is growing by more than 200% per year. Twitter, while having a "mere" 7 million users, has shown year-to-year growth of over 1000%. Using these types of tools, it has never been easier for people to track a company's sustainability performance and to widely disseminate their perspectives on it. We have entered an era of "radical transparency."

CHAPTER 13

The Stakeholder Perspective

13.1. GOVERNMENT

Governments in creating and developed nations are updating strategies in light of key sustainability issues, including ozone-depleting substance discharges, dangerous chemicals, water utilization, work, and human rights.

In the United States, there is restored enthusiasm for growing more compelling oversight of and responsibility for corporate exercises that effect society and the earth. Specifically, the eagerly awaited control of carbon outflows will give a more level playing field for bring down producers. However, these measures are very much dependent on the elections of 2017.

13.2. INVESTORS

More investors are currently requesting that organizations detail and measure sustainability dangers and opportunities in their monetary divulgences. As organization proprietors, they look to sustainability execution as a marker of solid administration, solid administration and long haul considering future development potential. Long-term speculators compensate organizations that coordinate sustainability into key objectives. Sustainability based stock files benchmark corporate execution against peer organizations and place weight on organizations to meet expanding desires.

13.3. WORKER'S UNIONS

Unions anticipate that organizations will address issues that influence employees, regardless of whether they are financial issues, for example, wages and social insurance, or natural issues, for example, wellbeing and environmental change. They effectively activate and speak to their individuals through formal arrangements, engagement on open approach, and coordinated effort with different businesses that help basic positions.

13.4. COMMON SOCIETY

NGOs and group bunches anticipate that organizations will address their ecological and social effects. Offering voice to planetary concerns and who and what is to come, and additionally those affected by corporate activity today, these gatherings weight organizations through legitimate activity, open crusades and cooperation to address a full scope of sustainability issues.

13.5. BUSINESS PARTNERS AND SUPPLIERS

Organizations expect that those they work with will take after similar principles that they improve the situation incorporating sustainability into their business. Business-to-business connections in this manner progressively consolidate sustainability models and criteria reflected through changes in Request for Proposals (RFP), and acquisition rules. As clients, organizations are themselves pushing sustainability crosswise over areas and the esteem chain. Gatherings of organizations in the car, clothing, electric utility, hardware, and pharmaceutical areas are among those working together to raise sustainability guidelines over their whole industry supply chains.

13.6. CONSUMERS

A developing section of individual customers are putting their cash where their qualities are, requesting a comprehension of the sustainability effects of the items and administrations that they purchase. Purchaser concerns incorporate the conditions under which items are made, the materials utilized, and post-utilize recyclability. An expanding enthusiasm for dependable sustainability data goes with this pattern, including at the purpose of procurement. The monetary emergency has not dulled buyer sense of duty regarding nature, with 78% of U.S. buyers in a 2009 Cone review showing that they are as likely or more inclined to purchase ecologically dependable items as they were before the emergency.

13.7. EMPLOYEES

One of the most grounded powers requesting change originates from inside.

Current employees and skilled employment competitors look for work that is important and of verifiable incentive to society, and they are set up to get a lower pay in return for work at a socially capable organization. They search out bosses that have an unmistakable vision for their commitment to an economical worldwide economy, and once inside, hope to impact the heading of corporate sustainability and drive upgrades through their particular obligations.

Similarly as with the current crumple of the monetary markets – originating from an absence of controls, disappointment of responsibility, and deficient exposure of dangers that speculators were expecting – sustainability and environmental change additionally exhibit broad concealed dangers to which the money related industry needs to give careful consideration.

Progressively, speculators are perceiving and making a move on administration issues, including natural and social issues that posture material business dangers and that can possibly specifically influence long haul monetary execution. On the other hand, they see significant advantages to be picked up by financing organizations that are proactive on these issues.

The Investor Network on Climate Risk (INCR) is centered around the monetary effects of environmental change. The gathering was built up in 2003 with an enrollment of twelve financial specialists speaking to $600 billion in aggregate resources, and now incorporates more than 80 speculators with $8 trillion in aggregate resources. INCR has been exceptionally dynamic, calling for solid national and worldwide atmosphere strategies, asking the U.S. Securities and Exchange Commission for institutionalized revelation of sustainability dangers, specifically captivating organizations, recording investor resolutions and making critical interests in clean innovation.

We have seen a blast of sustainability and particularly atmosphere centered research and speculation items in the previous couple of years, quite a bit of it occurring amidst one of our most noteworthy financial moderate downs. We have additionally observed the number of money-related records that address sustainability widen and extend.

A considerable lot of these records are being overseen by "standard" money-related houses, including the S&P/IFCI Carbon Efficient Index, HSBC Climate Index, Prudential Green Commodities Index, FTSE4Good, and NASDAQ's Global Sustainability 50 Index. Also, speculators are seeing the outcomes. For instance, KLD Research and Analytics (now part of Risk Metrics) propelled a Global Climate 100 Index and posted a 57% return (17% annualized) since its dispatch in 2006.

Notwithstanding remain solitary sustainability lists, money related businesses are incorporating sustainability into center basic leadership. Various banks, including Citi, Morgan Stanley, and Credit Suisse, have set up a carbon-centered due ingenuity process for any future loaning for coal-let go control and other carbon concentrated undertakings. Bank of America has likewise indicated initiative by setting a particular focus to diminish the rate of ozone-depleting substance outflows in its loaning to the utility business, and by revealing openly that it is utilizing a $20 to $40 per ton cost of carbon in assessing credits.

Socially dependable contributing has advanced from being a "boutique" issue, to developing at a quicker pace than the universe of all venture resources under expert administration – growing 18% in the vicinity of 2005 and 2007 – while professionally oversaw resources became under 3%. A report by Robeco Investment Management and Booz and Company anticipates that dependable contributing will develop to as much as 20% of aggregate resources under administration by 2015.

The budgetary emergency has additionally brought about an expanded concentrate on the hazard administration forms at money related organizations, with a push for more noteworthy straightforwardness, all-encompassing danger distinguishing proof and a more drawn out term center. The 658 signatories to the Principles for Responsible Investment, which all in all speak to over $18 trillion of advantages, trust it is in the long haul enthusiasm of their recipients to figure sustainability speculation basic leadership.

A 2009 Report by the Asset Management working gathering of the UNEP Financial Initiative declares that speculators have a guardian obligation to think about sustainability segments and those that don't may confront genuine hazard by opening themselves up to lawful liabilities.

The primary concern for business is this: financial specialists are compensating organizations that see how sustainability issues affect their business and that are executing procedures and activities that will empower them to flourish in a sustainable worldwide economy.

CHAPTER 14

The Eight Key Expectations

14.1. GOVERNANCE FOR SUSTAINABILITY

- **Board Oversight:** The Board of Directors will provide oversight and accountability for corporate sustainability strategy and performance. A committee of the board will assume specific responsibility for sustainability oversight within its charter.
- **Management Accountability:** The CEO and company management–from C-Suite executives to business unit and functional heads–will be responsible for achieving sustainability goals.
- **Executive Compensation:** Sustainability performance results are a core component of compensation packages and incentive plans for all executives.
- **Corporate Policies and System:** Companies will embed sustainability considerations into corporate policies and risk management systems to guide day-to-day decision-making.
- **Public Policy:** Companies will clearly state their position on relevant sustainability public policy issues. Any lobbying will be done transparently and in a manner consistent with sustainability commitments and strategies.

14.2. STAKEHOLDER ENGAGEMENT

- **Focus Engagement Activities:** Companies will systematically identify a diverse group of stakeholders and regularly engage with them on sustainability risks and opportunities, including materiality analysis.

- **Stakeholder Dialogue:** Companies will engage stakeholders in a manner that is ongoing, in-depth, timely, and involves all appropriate parts of the business. Companies will disclose how they are incorporating stakeholder input into corporate strategy and business decision- making.
- **Investor Engagement:** Companies will address specific sustainability risks and opportunities during annual meetings, analyst calls, and other investor communications.
- **C Level Engagement:** Senior executives will participate in stakeholder engagement processes to inform strategy, risk management and enterprise-wide decision-making.

14.3. DISCLOSURE

- **Standards for Disclosure:** Companies will disclose all relevant sustainability information using the Global Reporting Initiative (GRI) Guidelines as well as additional sector-relevant indicators.
- **Disclose in Financial Findings:** Companies will disclose material sustainability issues in financial filings.
- **Scope and Content:** Companies will regularly disclose significant performance data and targets relating to their global direct operations, subsidiaries, joint ventures, products, and supply chain. Disclosure will be balanced, covering challenges as well as positive impacts.
- **Vehicles for Disclosure:** Companies will release sustainability information through a range of disclosure vehicles, including stand-alone reports, annual reports, financial filings, websites, and social media.
- **Product Transparency:** Companies will provide verified and standardized sustainability performance information about their products at point of sale and through other publicly available channels.
- **Verification and Assurance:** Companies will verify key sustainability performance data to ensure valid results and will have their disclosures reviewed by an independent, credible third party

14.4. PERFORMANCE

1. **Operations:** Companies will invest the necessary resources to achieve environmental neutrality and to demonstrate respect for human rights in their operations. Companies will measure and improve performance related to GHG emissions, energy efficiency, facilities and buildings, water, waste, and human rights.
   a. *Greenhouse Gas Emissions and Energy Efficiency:* Companies will reduce greenhouse gas emissions by 25% from their 2005 baseline by 2020: Improving energy efficiency of operations by at least 50%. Reducing electricity demand by at least 15%. Obtaining at least 30% of energy from renewable sources.
   b. *Facilities and Buildings:* Companies will ensure that at least 50% of their owned or leased facilities, and all new construction, will meet rigorous green buildings standards. When setting up facilities, companies will follow best practices that incorporate sustainable land-use and smart growth considerations.
   c. *Water Management:* Companies will assess water-related impacts and risks and will set targets to improve water use and wastewater discharge, with priority given to operations in water-stressed regions.
   d. *Eliminate Waste:* Companies will design (or redesign, as appropriate) manufacturing and business processes as closed-loop systems, reducing toxic air emissions and hazardous and nonhazardous waste to zero.
   e. *Human Rights:* Companies will regularly assess key risks related to human rights throughout their entire operations, and will employ management systems that are aligned with internal policies and support the implementation of universal standards.

14.5. SUPPLY CHAIN

Companies will require their suppliers to meet the same environmental and social standards as the company has established for itself. Companies will establish sustainable procurement criteria, catalyze improved supplier performance, and facilitate disclosure of suppliers' sustainability information.

1. **Policies and Codes:** Companies will set supply chain policies

and codes aligned with overall social and environmental standards.

2. **Align Procurement Practices:** Companies will address sustainability performance in procurement criteria and contracting.
3. **Engaging Suppliers:** Companies will ensure that at least 75% of the company's Tier 1 and Tier 2 suppliers and 50% of Tier 3 suppliers meet the company's standards for sustainability performance.
4. **Measurement and Disclosure:** Companies will disclose a list of their Tier 1 and 2 suppliers and measure and disclose suppliers' sustainability performance.

14.6. TRANSPORT AND LOGISTICS

Companies will systematically minimize their sustainability impact by enhancing the resiliency of their logistics. Companies will prioritize low impact transportation systems and modes, and address business travel and commuting.

1. **Transportation Management:** Companies will develop transportation criteria that incorporate distance requirements from site to market and establish decentralized and localized distribution networks.
2. **Transportation Modes:** Companies will review logistics to prioritize low-impact transportation modes. 3. Business Travel and Commuting: Companies will decrease greenhouse gas emissions from business travel and employee commuting by 50% from a baseline of2005.

14.7. PRODUCTS AND SERVICES

Companies will design and deliver products and services that are aligned with sustainability goals by innovating business models, allocating R&D spend, designing for sustainability, communicating the impacts of products and services, reviewing marketing practices and advancing strategic collaborations.

1. **Business Model Innovation:** Companies will innovate business models to reduce material inputs and prioritize a transition to sustainable products and services.
2. **R&D and Capital Investment:** Companies will use sustainability as a primary filter through which all R&D and capital

investments are made. 50% of the R&D investment will be focused on developing sustainability solutions.

3. **Design for Sustainability:** Companies will approach all product development and product management decisions with full consideration of the social and environmental impacts of the product throughout its lifecycle.
4. **Marketing Practices:** Companies will align their marketing practices and product revenue targets with their sustainability goals, and will market their designed-for- sustainability products and services with at least the same effort as their marketing of other products.
5. **Strategic Collaborations:** Companies will collaborate within and across sectors to innovate and scale sustainable products and services, and contribute to the development of open source solutions.

14.8. EMPLOYEES

Companies will make sustainability considerations a core part of recruitment, compensation, and training, and will encourage sustainable lifestyle choices.

- **Recruitment and Retention:** Companies will incorporate sustainability criteria into recruitment protocols, employee performance processes, compensation, and incentives.
- **Training and Support:** Companies will develop and implement formal training on key sustainability issues for all executives and employees, and facilitate coaching, mentoring and networks for sustainability knowledge sharing.
- **Promoting Sustainable Lifestyles:** Companies will promote sustainable lifestyle choices across their community of employees through education and innovative employee benefit options.

CHAPTER 15

Climate Change: Present Urgency

The undeniably serious impacts of the ascent in worldwide temperature are being felt wherever on the planet through extraordinary climate occasions and cataclysmic events, filling in as a reminder to the effects of environmental change. For the individuals who are happy to tune in. Logical investigations have exhibited that we are balanced on the cusp of overall catastrophe, and the worldwide network is ending up progressively mindful of the looming emergency. Many are requesting activity; the world's countries and pioneers are investigating new innovations for feasible energy generation to anticipate a worldwide calamity.

As indicated by the Intergovernmental Panel on Climate Change (IPCC), a body committed to giving a target see on environmental change, 2018 is a tipping point for the worldwide condition. In 2018, we required an urgent year for atmosphere activity. Here are six noteworthy environmental change topics that rose amid the year.

15.1. THE ENVIRONMENT: EXTREME WEATHER EVENTS

Climate occasions and ecological calamities are dictated by an assortment of complex elements, both characteristic and man-made. It is difficult to property any single climate wonder legitimately to environmental change. Science makes it conceivable to discover how much environmental change impacts climate occasions, be that as it may. Logical examination uncovers that the ascent in worldwide temperature is in excess of a minor part of cataclysmic events and extraordinary climate occasions. Much of the time, it fills in as a basic factor.

About each area on the planet experienced extraordinary climate and cataclysmic events in 2018. The sea tempests of Florence and Michael caused monstrous annihilation in the southeast U.S. amid the Atlantic tempest season. In the Pacific, Super Typhoons Mangkhut and Yutu shook the Philippines, Guam, South China, and the Mariana Islands. Argentina and Uruguay experienced serious dry spell. The U.S. territory of California encountered the deadliest and most dangerous out of control fires on record and Europe has seen limits in both cold and warmth this year.

Overwhelming precipitation added to flooding and avalanches in Africa, India, Japan, Korea and the Caribbean islands of Grenada and Trinidad in 2018. A large number of individuals were dislodged and there has been across the board annihilation of homes and foundation. Flood conditions have exacerbated flare-ups of cholera in East Africa and leptospirosis ("rodent fever") in India.

In mid-December, the Center of Research on the Epidemiology of Disasters detailed that roughly 5,000 individuals have kicked the bucket and 28.9 million required crises help overall due to extraordinary climate. Research shows that financial misfortunes from atmosphere related calamities have expanded 151% in the previous 20 years, a pattern that is just expected to decline.

15.2. THE SCIENCE: UNEQUIVOCAL EVIDENCE

More researchers than any time in recent memory are directing concentration toward considering the effects of environmental change.

This year saw new examinations that have expanded our comprehension of the outcomes of the ascent in worldwide temperature. In October, the UN IPCC discharged a milestone report uncovering that worldwide temperatures are moving towards a cataclysmic 3°C ascend amid this century. The investigation supports quick and extraordinary changes to diminish worldwide temperature increment to at any rate 1.5°C.

Additionally discharged in October, ahead of time of the COP24 United Nations (UN) Climate Change Conference, the UN Environmental Emissions Gap Report clarified that current Nationally Determined Contributions (NDCs) are insufficient to connect the outflows hole by the deadline of 2030. The report declares that pressing activity is required to deflect the disastrous results of a determined ascent in worldwide temperatures.

Global NASA-drove studies have been concentrating Arctic ice misfortune and its effect on chilly movement, ocean level and floats in the Earth's turn hub. The aftereffects of these examinations clarify that the planet-wide impacts of environmental change are significant.

Research has likewise shown that the planet is amidst an eradication emergency. Earth is at present encountering the most noteworthy species cease to exist since the occasion that caused the annihilation of the dinosaurs 65 million years prior. While past termination occasions were brought about by such things as space rock strikes and characteristic moves in ecological conditions, the recent development is being driven by human exercises.

With the developing earnestness behind the science, global activists assembled in London to attract regard for this biological and natural crisis. Termination Rebellion (XR) protestors called upon the administration, private organizations and the media to perceive the desperate impacts of environmental change on the planet's species.

15.3. POLITICS AND SOCIETY: ACTION AND INACTION

The Extinction Rebellion was just a single development of numerous this years that looked to challenge an absence of atmosphere activity. Gatherings, for example, The People's Climate Movement, Rise for Climate and Strike 4 Climate Action assembled to request a suitable reaction to the emergency confronting our planet.

Another eminent occasion was the March for Climate, which occurred amid COP24, the greatest environmental change meeting since the Paris Agreement of 2015. Despite the way that the walk happened as a major aspect of a worldwide environmental change meeting, tranquil protestors were welcomed with harsh Polish law authorization nearness.

The IPCC and UN discharges whole reports, joined with COP24, left the subject of the ascent in worldwide temperature at the bleeding edge of universal discourse.

Logical proof has had the effect and looming outcomes of environmental change clear. Far-reaching worldwide responsibility to making unequivocal move even with this emergency has stayed lazy, be that as it may. The host nation of COP24, Poland, sent blended messages by putting coal at the focal point of the meeting. One of the world's driving producers, the U.S., has situated itself to pull back from the Paris Agreement at the most punctual conceivable chance.

Referring to the model being set by the Trump organization in the U.S., different countries are declining to participate with the universal network in environmental change endeavors. Brazil, the underlying planned host nation of COP25, has pulled back from that dedication. In the interim, Kuwait, Russia, Saudi Arabia joined the U.S. in declining to respect the milestone IPCC consider into the COP24 arrangements.

Approaching individuals from the U.S. Congress are dismissing the position of the current Presidential organization, nonetheless. Advocates of green innovations and supportability activities are supporting a Green New Deal, an arrangement to modernize foundation and put intensely in spotless and sustainable power source creation. While depicted as a "radical" plan by those with connections to the non-renewable energy source industry, "extraordinary" measures are called for by the U.S. to turn away worldwide debacle.

15.4. TECHNOLOGY AND INNOVATION: TRENDS TO WATCH IN 2019

Where there has been foot-delaying the political front, trendsetters and privately owned businesses are reacting to require the advancement and usage of new innovations. Advancement is being made with energy stockpiling and microgrid frameworks that use artificial intelligence (AI), Internet of Things (IoT) and blockchain to improve proficiency and financing openings.

In many creating nations, sustainable power source gives a chance to improve personal satisfaction and adjust to the effects of environmental change. Little-scale sun-powered home frameworks furnish a reasonable alternative in nations with low access to power. With the cost of sustainable power source proceeding to fall, the financial advantages of sustainable power source are obvious from Botswana to Boston. These mechanical advancements are probably the most significant things to watch in 2019.

15.5. TAKING ACTION: DEVELOPING AND IMPLEMENTING SOLUTIONS

On the bleeding edges of inclination the impacts of environmental change, it is nothing unexpected that little island creating states (SIDS) in the Caribbean have taken a place of administration in actualizing new energy and manageability activities. Planning to turn into the world's first-atmosphere shrewd zone, 26 nations and in excess of 40 private division accomplices have made an alliance to quick track atmosphere activity in the Caribbean area.

Key associations and submitted financing for atmosphere savvy speculations are being devoted to diminishing emanations and atmosphere related perils while supporting solid biological systems and verifying sustainable power source creation. This atmosphere shrewd zone will lessen helplessness with the impacts of environmental change while additionally fabricating monetary security for its natives.

Around the globe, others are additionally turning their psyches and hands towards creating answers for the difficulties displayed by environmental change.

The Nobel Economic Sciences Prize Committee exhibited the multi-disciplinary significance of the subject with its 2018 honor. The prize this year went to financial experts William Nordhaus and Paul Romer, who adjusted the western monetary development model to concentrate on natural issues. This acknowledgment of the critical issue of environmental change and the requirement for multi-disciplinary ways to deal with tending to it is noteworthy.

Other atmosphere activity pioneers have been perceived for the current year for endeavors to lessen discharges and advance supportability in everything from school cafeterias to football clubs. Where activity might be moderate with respect to governments and universal gatherings, obviously people, scholastics and privately owned businesses are venturing up to give true answers for location the test of environmental change.

15.6. INCREASING PUBLIC EDUCATION AND AWARENESS

Mindfulness and instruction about environmental change matters expanded for the current year. Writers and open figures utilized stimulation media and the web to attract consideration regarding the issue of environmental change and the need to seek after practical measures in every aspect of society.

English author, naturalist and supporter David Attenborough pointed out the worldwide issue of plastics contamination through his inclusion in the BBC's Blue Planet arrangement. Attenborough additionally played a basic job in COP24's People's Seat activity, conveying a moving discourse that characterized environmental change "our most prominent danger in a large number of years."

Despite what has to a great extent been an environmental change media power outage in the U.S., in excess of 100 meteorologists wore coordinating ties amid their late spring solstice communicates to exhibit mindfulness and solidarity. In Zambia, comic craftsman Mwelwa Musonko made and propelled another comic arrangement to bring issues to light about environmental change.

While trying to awaken those most safe, researchers detailed for the current year that environmental change stands to create an overall larger deficiency. As brew is made with grain, an especially atmosphere touchy harvest, the proceeded with ascend in worldwide temperature will inevitably make the refreshment uncommon and costly. In the event that extraordinary climate occasions, mass annihilations and the wobbling of the Earth itself aren't sufficiently convincing without anyone else, maybe a risk to the world's most prominent mixed drink will motivate more individuals to make a move.

CHAPTER 16

Renewable Energy Technology: Trends

16.1. ENERGY STORAGE

Energy storage assumes a significant job in adjusting force free market activity, and is vital to handling the discontinuity issues of sustainable power source. Matching a capacity framework with a sustainable power source guarantees a smooth and relentless power supply, notwithstanding when climate conditions are not ideal for energy age.

Batteries are the most well-known capacity gadgets utilized in sustainable power source frameworks and their utilization is expanding on both the private and matrix wide scale. Energy storage advances are relied upon to keep on improving, making their utilization increasingly reasonable and moderate. It is anticipated that capacity will speak to a center segment of all new energy innovations moving into the future, as both utility-scale and household energy stockpiling arrangements become more cost focused, dissolving the benefits of customary energy sources.

On the Caribbean island of Barbados, old electric vehicle batteries are being reused to give framework energy storage and broaden their helpful life expectancy.

16.2. MICROGRIDS AND AI

Microgrids are nearby energy frameworks that can work either independently or while associated with a bigger conventional lattice. They give energy autonomy, productivity, and assurance amid crises. Utilizing the abilities of artificial intelligence (AI) with microgrid controllers considers consistent adjustment and improvement of activity.

Arrangement of microgrids is ending up essentially snappier, with new programming taking into account plans to be finished in a solitary day. San DIEGO Tech Organization XENDEE is revealing a progressed toolbox for microgrid plan in association with the Worley Parsons Group to make accessible turnkey arrangements in up to a 90% less time and cost than ordinary techniques.

From California to the Pacific Islands, microgrids are becoming the dominant focal point. Palau is a Pacific island country and the thirteenth littlest nation on the planet, however, has focused on structure what might be the world's biggest microgrid – 35 MW of sun based boards matched with 45 megawatt-long stretches of energy stockpiling. This exertion is a piece of the country's objective to progress to 70% sustainable power source by 2050, perceiving that islands are on the cutting edges of encountering the impacts of environmental change and remain as ground zero for the worldwide energy progress.

16.3. ENERGY BLOCKCHAIN AND IOT

Initially created to record cryptographic money exchanges, blockchain innovation is being adjusted for use in the energy showcase. Blockchain is a morally sound computerized record that behaviors and records exchanges through a distributed system. The absence of centralization in blockchain leaves it as perfect for disposing of the brokers of power providers. It decreases energy disparity and wastefulness and enables purchasers to purchase and sell energy from different customers legitimately.

Matching the dispersed record innovation of blockchain with the ordinary gadgets that we use to get and pass on data, presently usually alluded to as the Internet of Things (IoT), and stands to profoundly affect energy frameworks. With the right applications, gadgets can self-governing purchase and sell energy at the ideal occasions, improve energy framework settings in a continuous setting and screen and break down execution of energy devouring gadgets.

These innovations are starting to be utilized for noteworthy ventures, for example, the Brooklyn Microgrid. The venture includes a network-controlled microgrid whose proprietors purchase and pitch capacity to one another by means of blockchain innovation. The accomplishments of undertakings like this guarantee they will be executed on a more extensive scale.

16.4. GRID PARITY AND FALLING COSTS

Framework equality happens when an elective energy can create control at an expense and execution level equivalent to or not as much as power produced from ordinary strategies. Sunlight based and wind has achieved equality in both cost and execution in numerous locales, and new advances keep on sharpening their aggressive edge.

The conventional model of huge, top-down and halfway-circulated energy creation is being supplanted by secluded, customer driven and equally dispersed power age. When thought to be hard to coordinate into the network, renewables are presently serving to fortify lattice unwavering quality and versatility. The use of blockchain, AI, and other mechanization advancements make renewables self-improving, expanding their productivity.

Sun-powered and wind is as of now more productive and savvy than regular sources, and developing advances will keep on improving their cost and execution. Joining the monetary advantages with a low ecological effect, we can hope to see renewables move from being a satisfactory energy source to a favored one.

16.5. BIG COMMITMENTS

An expanding number of organizations, urban communities and nations are grasping emanations decrease targets and atmosphere activity intends to meet the objective of restricting the ascent in worldwide temperature.

Until this point in time, more than 100 urban communities overall report that at any rate 70% of their energy generation is from renewables, and more than 40 are at present working on 100% inexhaustible power. Hundreds more have vowed to move in the direction of the objective of energy age that is 100% dependent on renewables. Perceiving their effect on environmental change, 158 organizations have likewise dedicated to progressing to 100% renewables.

Huge numbers of these enterprises and districts were enlivened to make these duties after the 2015 Paris Agreement. With the new data given by the IPCC's 2018 report, we can hope to see an expansion in intense duties to change from non-renewable energy sources to sustainable power sources.

16.6. ENERGY ACCESS ADVANCES IN DEVELOPING COUNTRIES

When looking at energizing new developments in energy innovation, it very well may be anything but difficult to overlook that a noteworthy level of the total populace stays without access to energy by any stretch of the imagination. A billion people live without power and several millions have problematic or restrictively costly energy sources.

Accomplishing all-inclusive access to energy is a basic part of tending to worldwide improvement challenges. The progressions that are happening in the energy market offer an answer for the issue of energy get to. Network-based microgrids could speak to the most practical method for conveying reasonable and solid capacity to those as of now living without it. Spotless, secluded and sustainable power source frameworks are perfect for a considerable lot of the networks that have been unfit to profit by regular incorporated types of energy age and conveyance.

While legislative issues regularly hinder advance, new energy advances and responsibilities from universal gatherings like the World Bank and SE for ALL are establishing a framework for giving energy access to creating nations. With their regularly improving expense and execution measurements, we can hope to see renewables progressively sent to give capacity to networks and areas that have for quite some time been left in obscurity.

CHAPTER 17

Eco-Efficiency

"This is what eco-efficiency is all about: combining the goals of business excellence and environmental excellence, and creating the link through which corporate behavior can support sustainable development."

—*Bjorn Stigson, President, WBCSD*

17.1. INTRODUCTION

Eco-efficiency is an administration theory that urges business to look for natural enhancements that yield parallel monetary advantages. It centers around business openings and enables organizations to turn out to be all the more earth dependable and increasingly gainful. It is a key business commitment to supportable social orders. As characterized by the World Business Council for Sustainable Development (WBCSD), "eco-proficiency is accomplished by the conveyance of intensely evaluated merchandise and ventures that fulfill human needs and bring personal satisfaction, while dynamically decreasing biological effects and asset force for the duration of the life-cycle to a level in any event in accordance with the Earth's assessed conveying limit." To put it plainly, it is worried about making more an incentive with less effect.

- Interface, one of the world's biggest makers of business floor covering, spared over $200 million from 1996 to 2002 through its maintainability endeavors.
- HP in California decreased its loss by 95% and spared $870,564 in 1998.
- STMicroelectronics, a Swiss-based innovation producer, spared

£38 million in vitality and $8 million in water costs, with a complete sparing over 10 years anticipated at $900 million.

- Dupont diminished vitality use by 33% at one office sparing over $17 million every year on power while decreasing ozone depleting substance contamination per pound of item significantly. In 2000, it spared nearly $400 million because of asset and profitability improvement.
- In five years, SC Johnson expanded generation by half while squander discharges were cut significantly, bringing about yearly cost investment funds of more than $125 million.
- United Technologies Corporation's locales disposed of just about 40,000 gallons for each time of waste water and spared over US$50,000 every year with a crucial change in the manner it deals with its test cells, underground stockpiling tanks and waste streams.

There have been incredible advances in the utilization of eco-efficiency standards to this present reality. Industry, for instance, has had impressive achievement in diminishing contamination and discharges, and killing dangerous materials from creation forms. Previously, business saw the earth and economical advancement as issues and hazard factors. Today, they are additionally observed as circumstances – wellsprings of productivity improvement and development. Eco-efficiency is particularly a piece of this image.

Fundamentally, it is tied in with accomplishing more with less: conveying more worthwhile utilizing less assets. In the event that you spare vitality, for instance, you cut your expenses while additionally decreasing unwelcome yields, for example, emanations. Eco-efficiency isn't constrained just to making gradual proficiency upgrades in existing practices and propensities. It ought to invigorate imagination and advancement in the quest for better approaches for getting things done. Nor is eco-efficiency restricted to regions inside an organization's limits, for example, in assembling and plant the board. It is additionally legitimate for exercises upstream and downstream of a producer's plant and includes the supply and item worth chains.

Thus, it very well may be an incredible test to advancement engineers, buyers, item portfolio directors, showcasing masters and even account and control. Organizations can utilize eco-efficiency as an indispensable social component in their strategy or statements of purpose. They can likewise set eco-proficiency targets for their natural or incorporated administration frameworks. Also, it is a valuable instrument for observing and detailing execution, and for helping the company's correspondence and exchange with its partners. Eco-proficiency openings can develop anytime in the whole life-cycle of an item. This implies workers need to comprehend what eco-proficiency is, the worth it can bring to an organization and how to get it going. This thusly requires building aptitudes and comprehension so as to coordinate eco-proficiency crosswise over business activities, segments, nations and issues, and permitting space for development and imagination.

The idea has moved from counteracting contamination in assembling ventures to turning into a driver for advancement and intensity. Organizations actualize eco-efficiency to advance their procedures, transform their losses into assets for different enterprises, and drive development that prompts items with new functionalities. Eco-efficiency is a functional methodology yet not a panacea. Furthermore, it will never function as an extra to a business – it must be a necessary piece of a system. Such a system will have a solid spotlight on mechanical and social development, responsibility and straightforwardness, just as on participation with different pieces of society with the end goal of accomplishing the set destinations.

Eco-efficiency can help organizations create and effectively execute a business procedure toward supportability just if it's in the hearts and psyches of workers. Exhibiting the estimation of an eco-effective methodology will enable workers to perceive why it is significant for the organization to actualize and inspire towards activity. Eco-efficiency has been illustrated, through several case models, to work for organizations all things considered, in every single modern division and in all areas. It is basic to draw on the scope of apparatuses, methodologies and models that as of now exist inside the pioneers in this field. Eco-proficiency additionally requires a scope of abilities and capacities from getting definitions and predicaments, dissecting partner points of view to undertaking an actual existence cycle appraisal, incorporating thinking crosswise over business tasks, coordinating and consulting with outer accomplices, and estimating and assessing sway. Eco-efficiency is, truth be told, work in advancement and will keep on being so in light of the fact that it is basically a dynamic as opposed to a static procedure.

The motivation behind the module is in this way to help organizations advocate and teach the eco-efficiency story up until this point:

- the business drivers and benefit of embracing eco-efficiency.

- the key methodologies and apparatuses for executing eco-proficiency.
- how to incorporate eco-efficiency into business basic leadership forms.

Eco-proficiency isn't adequate without anyone else's input since it coordinates just two of supportability's three components, financial aspects and biology, while leaving the third, social advancement, outside its grip. By pushing and instructing eco-effectiveness, representatives can more readily comprehend the effects of the organization and its association with society. This will empower business to look past eco-efficiency so as to acquire its permit to work, advance and develop.

17.2. WHAT IS ECO-EFFICIENCY?

At a fundamental level, eco-proficiency signifies, "accomplishing more with less." It is an administration procedure that joins ecological and financial execution. It empowers progressively effective generation forms and the formation of better items and administrations while diminishing asset use, waste, and contamination along the whole worth chain. It makes more an incentive with less effect by unlinking merchandise and ventures from the utilization of nature. Not exclusively would it be able to spare creation costs yet can likewise open up new wellsprings of income for organizations.

These are a couple of compact meanings of the idea:

- The WBCSD characterizes eco-proficiency as being "accomplished by the conveyance of intensely evaluated merchandise and ventures that fulfill human needs and bring personal satisfaction, while logically decreasing natural effects and asset power for the duration of the life-cycle, to a level at any rate in accordance with the Earth's assessed conveying limit."
- The OECD has called eco-efficiency "the effectiveness with which biological assets are utilized to address human issues" and characterizes it as a proportion of a yield (the estimation of items and administrations created by a firm, segment, or economy all in all) partitioned by the info (the total of ecological weights produced by the firm, the area, or the economy).
- Academic specialists and professionals term eco-proficiency the blend of "monetary and natural productivity in parallel."
- The European Environment Agency (EEA) characterizes it

as "more welfare from less nature" and says it comes through decoupling asset use and toxin discharge from financial advancement and by and large welfare (see Figure 17.1).

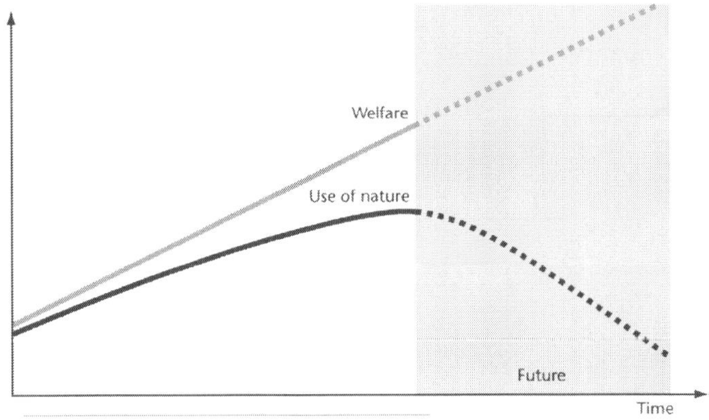

Figure 17.1: Greater Welfare from Less Nature. Source: WBCSD.

17.3. KEY DRIVERS

In spite of the fact that the expression "eco-effectiveness" has been being used for over 10 years, it has just been grasped by the more extensive business network in the recent years. How did this occur? Whatever degree did the administrative condition push business towards progressively effective practices? Did organizations react to client needs or perceive the upper hands in an eco-proficient methodology? Or on the other hand does maintainability basically bode well? The accompanying timetable recognizes key occasions and drivers towards the rise, acknowledgment, and execution of eco-effectiveness.

- 1962 – Silent Spring: Publication of Rachel Carson's fiery book on the synthetic business, which featured significant inquiries regarding mankind's effect on nature. This book brought issues to light of the job of business and gave a platform to the ecological development.
- 1972 – Limits to Growth: This production issued an unmistakable cautioning that monetary development was spending assets at an unsustainable rate and the points of confinement to development of the planet would be come to inside the following 100 years. Testing standard way of thinking of the time, the Club of Rome

anticipated this would bring about an abrupt and wild decrease in both populace and modern limit. While the investigation was very censured, it in any case started genuine exchange on the Earth's conveying limit and featured the likelihood to adjust these development slants so as to set up a state of natural and monetary security that is feasible far into what's to come.

- 1970s – Pollution Anticipation: Pollution counteractive action moved accentuation from "controlling" contamination once it has been made to "forestalling" its creation in any case. The US producer 3M was instrumental in building up the idea with its "Contamination Prevention Pays," a program executed in 1975 to dispose of contamination at source. They accomplished more than $800 million in combined first-year reserve funds.
- 1972 – UN Conference on the Human Environment: Leaders from industrialized and creating countries gathered at the UN Conference on the Human Environment in Stockholm, Sweden, so as to portray the "rights" to a solid and beneficial condition. Around then the spotlight for organizations was to a great extent on ecological assurance by following enactment.
- 1980s – Environmental Development: The ecological development developed as buyers turned out to be increasingly aware of the natural effects of the items they were purchasing. In 1984, a toxin gas hole happened at a Union Carbide pesticide processing plant in Bhopal, India, and the jobs and duties of business were addressed by society.
- 1987 – The Bruntland Report, Our Common Future: This fundamental report from the World Commission on Environment and Development inspected the world's ecological and advancement difficulties and how to illuminate them. An invitation to take action, the book recommended connection among governments and the requirement for a typical methodology.
- 1989 – Cleaner Creation: The United Nations Environment Program (UNEP) formalized the idea of cleaner generation, and characterized it as "the ceaseless utilization of an incorporated preventive natural technique connected to procedures, items and administrations to increment eco-proficiency and decrease dangers to people and the earth." This center turned out to be fairly increasingly proactive as business grasped the idea of

contamination aversion by means of cleaner generation, not really guideline driven control. Governments started to work more agreeably with organizations as opposed to by order and-control guideline.

- 1991 – The WBCSD: The Business Council for Sustainable Development (BCSD) – presently the WBCSD – first utilized the term eco-proficiency in 1991. This idea set up the connection between ecological execution and monetary advantages. Stephan Schmidheiny and the BCSD took the idea worldwide through a production entitled Changing Course. Having instituted the saying eco-proficiency, the WBCSD proceeded to have various exchanges, workshops and productions.
- 1990s – Setting Eco-Efficiency: Eco-efficiency focuses on the constrained physical conveying limit of the Earth turned out to be increasingly acknowledged with clear targets, and progressively solid arrangements: Eco-efficiency and development as devices to unlink financial development from physical development.
- The Factor 10 Club, as a universal assemblage of senior government, nongovernment, industry, and scholarly pioneers required the selection of a calculate ten increment vitality and asset profitability in created nations, while lessening the complete utilization of normal assets internationally. It requested a flat out unlinking of financial development from the utilization of regular assets.
- 1992 – UN Conference on Environment and Development: Rio Heads of state met for the primary global Earth Summit to address critical issues of natural insurance and financial advancement. Motivation 21, an arrangement for accomplishing reasonable advancement in the 21st century, was embraced.
- The expression "supportable improvement" thought about expanding center the advancement motivation and was grasped through crafted by the World Commission on Environment and Development (Our Common Future in 1987). Notwithstanding eco-efficiency, organizations started to address corporate social duty (CSR) and how organization tasks influence society as a rule.
- 1998 – Factor Four: "Factor four" was presented in 1998 by the Rocky Mountain Institute. It alludes to a theoretical fourfold

increment in "asset profitability," achieved by at the same time multiplying riches and dividing asset utilization. It proceeds to outline advances that can convey the essential upgrades.

- 2002 – World Summit The full execution of Agenda 21: The Program for Further Implementation of Agenda 21 and the Commitments to the Rio standards, were reaffirmed at the World Summit on Sustainable Development (WSSD) held in Johannesburg, South Africa, in 2002.

17.4. THE BUSINESS CASE

The business basis for eco-efficiency is clear: it bodes well. Eco-efficiency improves business and ecological execution, and enables organizations to get out before market and administrative patterns, to diminish costs, to increase upper hand and to guarantee long haul gainfulness and manageability.

Being productive is consistently a high need for each organization. Be that as it may, on the off chance that it incorporates making monetary worth and diminishing ecological effect and asset use simultaneously, the worth included turns out to be considerably progressively huge. The business case for eco-efficiency applies to each region of action inside an organization – from dispensing with dangers and finding extra reserve funds to recognizing new market openings. Five parts of eco-efficiency have been distinguished that make it a fundamental key component in business (see Figure 17.2).

Scaling up eco-proficiency needs individuals who completely comprehend the idea and up eco-efficiency its potential. It ought not be restricted essentially to making gradual effectiveness enhancements in existing practices and propensities. Or maybe, it ought to animate imagination and development in the quest for better approaches for getting things done. Second, it isn't constrained to regions inside an organization's limits, for example, assembling and plant the board, yet is similarly as legitimate for the whole supply and utilization generation esteem chains. Eco-proficiency's adaptability makes it helpful to all organizations through evolving times. The necessities of clients change, thus does their comprehension of the earth and the degree of eco-limit. New dangers develop, for example, environmental change, and existing ones become better described. The proceeding with weight of financial and populace development implies that a few difficulties –, for example, the accessibility of clean water – become progressively genuine. That is even more motivation to concentrate on eco-efficiency as an adventure instead of a goal, a procedure as opposed to a panacea.

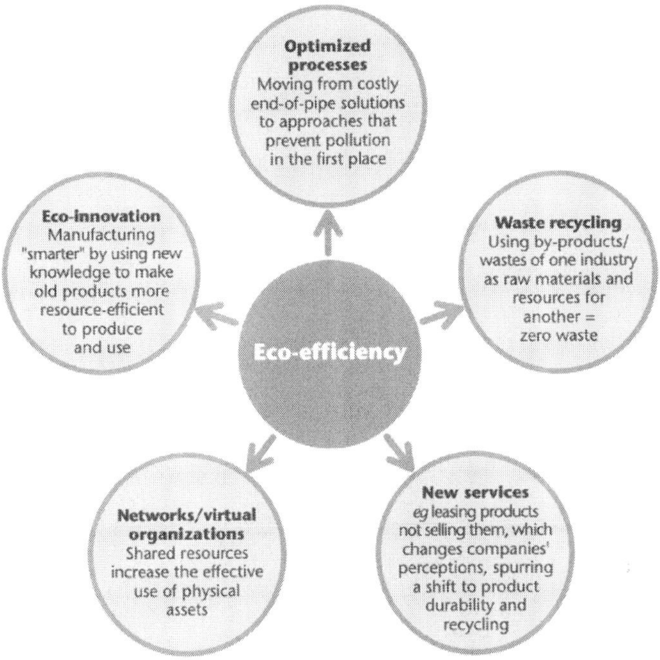

Figure 17.2: The business case for eco-efficiency. Source: WBCSD.

Money related markets have additionally begun to take a gander at parts of natural execution. Forward-looking experts are choosing manageability pioneers since they perceive that organizations which have executed eco-efficiency as a business idea outflank their rivals essentially. From a macroeconomic viewpoint, eco-efficiency encourages wealthier nations to develop more subjectively than quantitatively – giving more administration, capacity, and worth, not changing more materials into vitality and waste. Eco-efficiency likewise enables creating nations to keep on developing quantitatively, yet with increasingly productive asset utilization and natural effect. Given that the idea is tied in with creating more with less (just as diminishing contamination), at that point poor nations remain to profit significantly, particularly poor nations where feedstocks, for example, oil are costly and some regular assets, for example, wood and water, are as of now hard to come by. Actually, improving productivity of activities might be the main manner by which many creating world organizations can contend in the worldwide market.

Figure 17.3 shows that nations with more grounded natural administrative systems are progressively focused all in all. This dissipates the hypothesis that there is a "race to the base," for example venture will stream to nations where ecological norms are less inflexible.

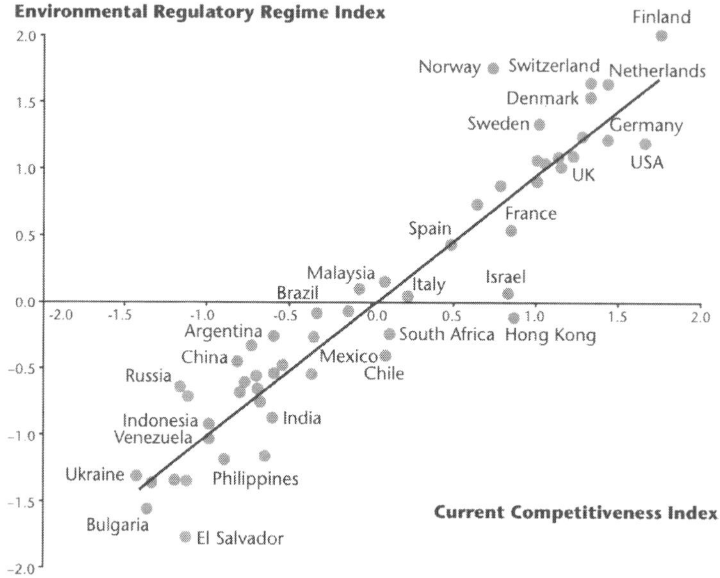

Figure 17.3: Environmental Regulatory Regime Index. Source: WBCSD.

17.5. IMPLEMENTING ECO-EFFICIENCY

Eco-proficiency can be executed along the whole worth chain of an item or administration, not simply inside the physical limits of the plant. For certain organizations, the principle unsafe consequences for the earth with which they are related really happen outside their wall – either upstream in the crude material age and provider preparing stages, or downstream in the item use or transfer stages. In light of this, eco-efficiency can be accomplished through seven key approaches:

- Lessen material force
- Vitality force limited
- Scattering of lethal substances is diminished
- Embrace reusing
- Benefit from utilization of renewables
- Broaden item strength
- Administration force is expanded

Figure 17.4 represents the four noteworthy zones that give chances to expanded eco-productivity. It can help organizations to distinguish where the open doors can be found and who, inside and outside the organization, will be associated with abusing them.

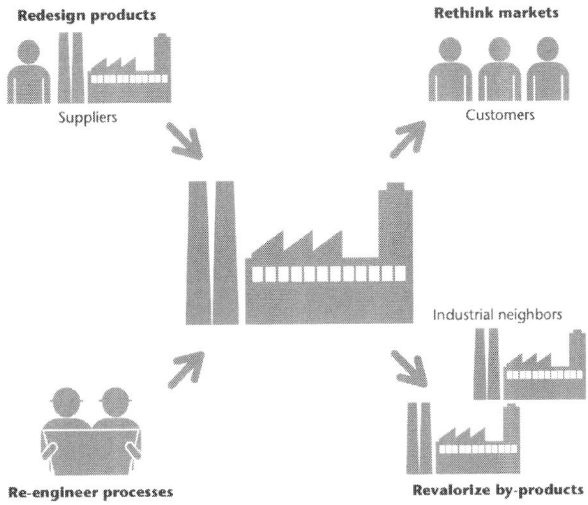

Figure 17.4: The four areas of opportunity for eco-efficiency. Source: WBCSD.

Opportunities to: Re-engineer processes	• Optimize processes and procedures to minimize resources (materials, time, etc.) • Maximize energy and water savings • Eliminate hazards where possible • Save through systematic risk minimization • In production, procurement, sales, R&D, human resources, etc.
Revalorize by-products	• New revenue sources from "waste" products • Cheaper resources from optimal by-products from others • Maximize use of purchased virgin resources • Benefits from cooperation with other companies
Redesign products	• Consider entire value chain in product design – supply chain through disposal • Make products upgradeable and recyclable • Create new and increased functionality
Rethink markets	• Know the customer • Sell functional (rather than material) offerings • Provide users with comprehensive solutions • Create new businesses with add-on services • Improve customers' eco-efficiency • Rethink opportunities in light of changing market conditions (new taxes, increased consumer awareness, etc.)

Figure 17.5: The four areas of opportunity for eco-efficiency in detail. Source: WBCSD.

All divisions inside an organization can add to expanding eco-effectiveness: tasks, acquisition, R&D, deals, advertising and the executives all have their part to play. Numerous organizations, perceiving this, have made eco-proficiency part of their general business system. They have understood that key eco-proficient open doors lie in their assembling, yet in addition along the whole store network just as in the utilization of their item and administrations. Eco-efficiency for them has turned into a noteworthy driver of advancement and advancement, a vehicle that causes them meet their monetary and ecological targets.

The WBCSD accepts that it is of fundamental significance that top administration become tied up with the idea of eco-effectiveness. Those enterprises where eco-proficiency is on the CEO's plan have had the option to gain genuine ground. Their presentation isn't restricted to a couple of gradual upgrades in determined angles. Rather, they start to jump eco-efficiency with imaginative items, new administrations and a focused business system toward manageability.

17.6. THE ECO-EFFICIENCY TOOL STASH

Actualizing eco-efficiency along the whole worth chain of an item or administration involves different methods and devices. Each instrument in the eco-efficiency tool kit keeps on developing. Eco-efficiency can be comprehended as a perspective (e.g., how would I lead business in a financially and ecologically productive way?) and an administration apparatus (e.g., efficiently applying the seven components to business basic leadership will improve my eco-proficiency). Usage of eco-efficiency can go from straightforward and to some degree standard mechanical practices identified with improving asset and vitality productivity to exceptionally inventive item and procedure update activities, in which natural or ecological contemplations are utilized as an impetus for change. To help this scope of uses there are numerous apparatuses accessible. Beneath, we have featured a couple of those most regularly utilized.

17.7. ENVIRONMENTAL MANAGEMENT SYSTEM (EMS)

An EMS is characterized as the authoritative structure, duties, rehearses, management methods, procedures, and assets for executing and dealing with an association's natural undertakings while guaranteeing adjustment to its strategies, models, and partners' desires. An EMS can help recognize where eco-efficiency openings happen inside an organization.

17.8. LIFE-CYCLE MANAGEMENT

The Society of Toxicology and Chemistry's Life-Cycle Management Working Group has characterized life-cycle the board as "an adaptable incorporated system of ideas, methods and techniques to address ecological, financial, mechanical and social parts of items and associations to accomplish persistent natural improvement from an actual existence cycle point of view." LCM can enable associations to coordinate or design eco-efficiency into business basic leadership.

17.9. LIFE-CYCLE ASSESSMENT

Life-cycle assessment is a basic leadership instrument to recognize natural weights and assess the ecological outcomes of an item, procedure or administration over its life-cycle from support to grave or support to support on account of items that can be recouped and reused or reused. By giving a framework wide perspective on an item, LCA is helping industry, governments, and research foundations decide the approaches to advance the eco-efficiency of a whole item framework rather than only a specific office or procedure.

17.10. ECO-INNOVATION, DESIGN FOR ENVIRONMENT (DFE) OR ECO-STRUCTURE

The reconciliation of ecological contemplations into item and procedure plan. Crucial to DfE is the utilization of apparatuses and practices that support or Eco-plan natural duty and at the same time decrease costs, advance intensity and improve development. Structure for Environment (DfE) enables associations to incorporate eco-efficiency right where it is most required – at the theoretical arranging and configuration organize.

17.11. NATURAL SUPPLY CHAIN MANAGEMENT (SCM)

Production network the executives (SCM) is the association of exercises to address the presentation of material, parts, merchandise and enterprises that an association purchases and employments. There is a developing pattern toward expanding ecological execution prerequisites down the store network from item producers. ESCM can enable associations to distinguish eco-efficiency openings in their production network.

17.12. CLEANER PRODUCTION

UNEP characterizes Cleaner Production as the nonstop utilization of an incorporated preventive natural system connected to procedures, items and administrations to increment eco-proficiency and lessen hazard for people and the earth. From multiple points of view Cleaner Production is tied in with operationalizing eco-efficiency at the procedure level (e.g., by preserving crude materials and vitality, killing poisonous crude materials and diminishing the amount and lethality all things considered and squanders before they leave a procedure) and the item level (e.g., by decreasing effects along the whole life-cycle of the item, from crude material extraction to a definitive transfer of the item). There are various different instruments and methodologies being utilized to help eco-efficiency.

They incorporate life-cycle costing, modern nature, green acquisition, contamination counteractive action and others. As a rule these apparatuses endeavor to decrease the ecological effects of organizations, and their items or administrations while all the while improving money related execution. The decision of which devices an association uses relies upon various components including their own inner approaches and limit, the nature of their connections with the indigenous habitat, the level of control they have over item plan and supply chains (e.g., an organization that does not plan items will have little use for eco-structure) and their specific market and administrative drivers.

17.13. ESTIMATING ECO-PROFICIENCY

"The familiar saying 'Just what completes estimated gets' is as valid for this situation as anyplace else," says Markus Lehni, chief of Global Sustainability Services at the board consultancy Deloitte and Touche. "Setting targets and estimating advancement is significant, both inside and remotely. With great measurements, we can all the more likely show how these things are associated naturally and economically."

Companies may gauge their eco-efficiency execution for various reasons – for instance:

- to track and record execution and advancement;

- to distinguish and organize open doors for development; and
- to recognize cost reserve funds and different advantages identified with eco-effectiveness.

An organization may likewise need to exhibit why improvement is restricted in specific regions or won't be conceivable to the degree expected by specific partners. Eco-efficiency pointers additionally help administrators settle on choices on an item or business portfolio. They can furnish administrators with the data on the most proficient method to make a business portfolio more eco-productive or progressively practical in general – and normally increasingly gainful too. Checking and detailing eco-efficiency freely is an approach to convey a key component of an enterprise's advancement on practical improvement to outside crowds, including financial specialists, safety net providers, purchasers, and other intrigued gatherings.

Basically, advance in eco-proficiency can be accomplished by giving more worth per unit of ecological impact or unit of asset expended:

$$\frac{\text{product or service value}}{\text{environmental influence}}$$

Utilizing this essential condition, organizations can ascertain eco-efficiency in various ways. The selection of markers will rely upon the requirements of individual leaders:

- A plant administrator may wish to concentrate on the quantity of items delivered per kilojoule of vitality devoured during assembling.
- A money related examiner may rather concentrate on the financial estimation of items sold per kilojoule.

Worth and natural impact can likewise be estimated for various elements, for example, generation lines, fabricating destinations, or whole partnerships, just as for single items, advertise fragments, or whole economies.

17.14. THE WBCSD STRUCTURE FOR ECO-EFFICIENCY DATA

Lehni calls attention to that the numbers themselves are not adequate to make the eco-efficiency case: "these figures need understanding, correlations with different organizations in a similar segment, and benchmarks. The majority of this makes validity among financial specialists and partners."

The WBCSD in this way has built up a typical system for eco-proficiency markers, with wording steady with the ISO 14000 arrangement and the Global Reporting Initiative (GRI). They characterize three degrees of association for eco-efficiency data – classifications, angles, and pointers.

- Categories are wide regions of ecological impact or business esteem. Where several perspectives are taken into account and classified accordingly.
- Indicators are the particular proportions of an individual angle that can be utilized to follow and show execution. A given perspective may have a few pointers.

The accompanying demonstrates the three classifications distinguished and their primary related viewpoints.

Product or service value	Environmental influence in product or service creation	Environmental influence in product or service use
• volume/mass • monetary • function	• energy consumption • materials consumption • natural resources consumption • unintended events • non-product output	• product/service characteristics • packaging waste • energy consumption • emissions during use/disposal

All three categories of eco-information must be:

- scientifically supportable;
- environmentally relevant;
- accurate; and
- useful for all kinds of businesses around the globe.

Every one of the three classifications of eco-data must be:

- logically supportable;
- earth significant;
- exact; and
- valuable for a wide range of organizations around the world.

A definitive point is to improve the business execution and screen it with measures that are straightforward and unquestionable, and in this manner important to business directors just as to outside partners. While the accentuation is on territories under direct administration control, we likewise perceive there are pertinent issues upstream (e.g., with providers) and downstream (e.g., in item use) of an organization's exercises. For the most part appropriate pointers can be utilized by for all intents and purposes all organizations.

Just as being pretty much all around significant, every one of these pointers identifies with a worldwide natural concern or business worth and strategies for estimation are built up and definitions acknowledged all around. Every other pointer, which don't meet these criteria, have been named business specific, implying that they are bound to be exclusively characterized starting with one business or one part then onto the next. These markers are not really less significant than the main gathering. That judgment will rely upon the idea of an individual business.

They are simply less generally relevant. An organization's eco-efficiency execution profile will incorporate the two sorts of markers. The pointers fall into two gatherings, in light of the eco-efficiency equation which unites the two eco-measurements of economy and biology to relate item or administration incentive to natural impact.

For the most part appropriate markers for item or administration worth are:

- amount of products or administrations delivered or gave to clients, and
- net deals.

Those identifying with the natural impact in item or administration creation are:

- vitality utilization,
- materials utilization,
- water utilization,
- ozone harming substance outflows, and
- ozone-exhausting substance emanations.

While the ecological impact of items or administrations being used is significant, we have not distinguished any commonly pertinent pointers for this classification of eco-proficiency. All pointers in this gathering are viewed as business-or item explicit. The accompanying extra pointers could turn out to be commonly material if current endeavors to create worldwide concurrence on estimation techniques are effective:

- extra money related worth markers,
- fermentation emanations to air, and

- all out waste.

Be that as it may, few for the most part appropriate pointers is useful in surveying organizations' eco-proficiency in light of the fact that a multiplication of measures would make it hard for reports all things considered and reasonable, particularly remotely. A little center of regular pointers will help learning and equivalence crosswise over time, divisions and enterprises. Simultaneously, singular organizations ought to build up a more full depiction of their exhibition by including business-explicit markers.

17.15. PROVIDING DETAILS REGARDING ECO-PROFICIENCY

There are significant viable issues for setting up an organization's eco-proficiency profile and detailing inside and to outside partners. These incorporate choosing limits for the information, where to discover information and how to assemble it properly, assessing affectability, mistake and issues of change and transformation. It is significant that organizations give some point of view on issues, for example, the degree and restrictions of their markers, with the goal that clients comprehend the idea of the data gave.

The inquiries of total and benchmarking are especially significant. While total will frequently be attractive, it might darken conceivably significant data about an endeavor's eco-efficiency execution. For instance, various kinds of inconsequential discharges, or emanations crosswise over various geographic areas, might be included, anticipating a substantial appraisal of potential effects on the earth. Essentially, eco-efficiency gauges from various items, forms, or geographic zones might be consolidated and arrived at the midpoint of, darkening insights concerning the presentation of individual units. Conglomeration should along these lines be finished with consideration, and straightforwardly, so the confinements of the data can be appropriately comprehended. This is particularly the situation when information is accounted for freely and examinations are made between various organizations, procedures, or items.

Outside clients of eco-efficiency data are probably going to need to make examinations among organizations and after some time – yet it is critical to perceive the inborn assorted variety of business and the specific conditions of individual organizations.

Examinations ought to be made just when the organizations being thought about give a similar item or administration. It is likewise critical to perceive that the item arrangement of various organizations frequently change, and this may influence eco-proficiency figures paying little mind to an association's ecological endeavors. The effect of such changes ought to be clarified by organizations in revealing eco-efficiency information, to enable clients to translate reports precisely. The culmination of this is organizations can utilize eco-proficiency data to build up their item blend and movement toward a more eco-productive portfolio.

Pattern information is significant with the goal that adjustments in execution after some time or contrasted with a reference point can be surveyed. Information can be exhibited as outright figures, eco-efficiency proportions, ordered to a chose year, or communicated with respect to an anticipated objective. Execution can likewise be communicated in respect to an industry normal, if that is accessible. On the off chance that lists are utilized, in any case, the outright numbers basic them ought to likewise be distributed with the goal that clients can compute their very own proportions. It is essential that information introduced in such diagrams be really similar so the patterns delineated precisely speak to eco-proficiency execution.

We propose the accompanying five components for a synopsis eco-efficiency profile, in addition to a confirmation proclamation if the report has been remotely checked on:

- Organization profile – to give a setting to the eco-efficiency data, including the quantity of workers, business sections, essential items and real changes in the structure of the organization.
- Value profile – markers from the "esteem" bit of the WBCSD system, including budgetary data, the amount of items, or practical pointers for explicit items.
- Environmental profile – including commonly material ecological impact markers just as business-explicit pointers identifying with item or administration creation and use.
- Eco-efficiency proportions – notwithstanding giving in the past two components the essential "numerator" and "denominator" information for assessing eco-proficiency, organizations may likewise wish to give computations of eco-efficiency pointers that they consider most important and significant for their business.
- Methodological data – covering the methodology used to

choose pointers, information accumulation procedures, and any restrictions on utilization of the information.

The WBCSD suggests that organizations incorporate eco-efficiency data into their general basic leadership and interchanges forms. Inside, it ought to be a piece of routine administration frameworks. Remotely, eco-efficiency markers could be given in professional workplace or supportability reports as one of the coordinating components between the three mainstays of manageability, and could be incorporated into existing budgetary reports as an expansion to unadulterated money related detailing.

17.16. ECO-EFFICIENCY AND GOVERNMENTS

Eco-efficiency works in organizations that attempt it; that is the uplifting news. The all the more alarming news is that eco-proficiency isn't yet being taken a stab at a scale that genuinely has any kind of effect in the manner most business is led today – or at a pace that will altogether stem the tide of harm to the common habitat as populaces develop and poor countries create. Venturing up endeavors in eco-efficiency will require another agreement among society, government, and business. Under it, corporate pioneers would promise to put resources into eco-proficient development – that is, to accomplish radical instead of gradual ecological improvement over the long haul, to work to decrease worldwide imbalances, and to be capable managers and network members.

Governments would consent to set up a system of guidelines, monetary instruments and data revelation to support long haul advancement and punish free riders without harming rivalry. Approach measures to use business activities for more eco-proficiency can incorporate components, for example,

- recognizing and disposing of unreasonable appropriations;
- disguising ecological expenses;
- moving duty from work and benefit to asset use and contamination;
- creating and executing financial instruments; and
- advancing deliberate activities and arranged understandings.

Governments can utilize different motivating forces to advance activity toward advancement and bolster activities to progress eco-efficiency – compensating the main edge organizations and putting weight on the slouches. Motivating forces to remunerate eco-efficiency will control development the correct way and make new items and administrations.

The EEA, built up by the European Union in 1990 to fill in as the center point of the European Environmental Information and Observation Network (EIONET), has received eco-proficiency proportion markers for nations, requesting an outright and relative "unlinking of development of welfare from the utilization of nature." It plans to gauge and contrast monetary areas and nations and each other as per their eco-efficiency status and enhancements.

The OECD condition services in 1996 watched: "A system to improve eco-efficiency may empower industry, governments, families to decouple contamination discharge and asset use from financial action" (Figure 17.5). The OECD researched the capability of the eco-efficiency idea "in the light of studies proposing that factor-of-ten proficiency enhancements are both fundamental and conceivable in the following 30 years," and suggested the utilization of eco-efficiency proportions as full scale level indicators.

The OECD distinguished development as the key driver for improving eco-efficiency and said that it was best animated by solid challenge, administrative motivators, a powerful procedure of scattering "best practice," and the nearness of a decent atmosphere for advancement.

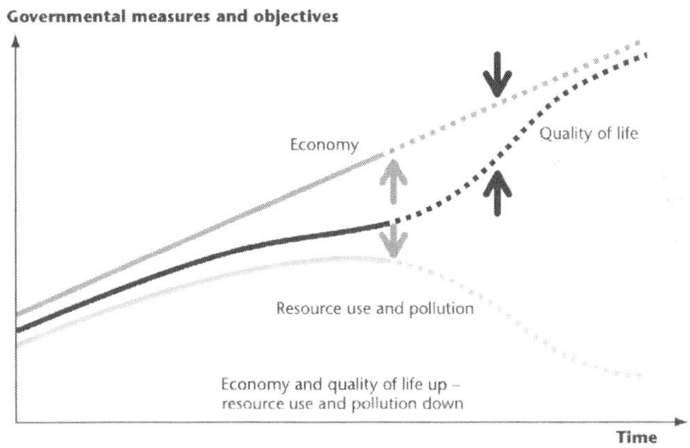

Figure 17.6: Decoupling economic growth form resources use and pollution. Source: WBCSD.

What is required presently are radical upgrades, better item structure, and further appropriation of eco-proficient standards by little and medium-sized undertakings, which contain most of industry in many nations. This calls for more noteworthy initiative by business, governments, and common society, further refining of targets and measurements, and persistent improvement.

It will likewise mean defeating obstructions to encourage extension of eco-proficiency, including the nonappearance of the correct market signals and inadequate administrative structures.

Regardless of these hindrances, eco-proficiency has increased enough energy to reclassify the manner in which many driving organizations lead their regular business and view their general surroundings. For these organizations, dematerialization, shut creation circles, administration expansion, and utilitarian augmentation have progressed toward becoming methods for working together more gainfully. The move from a supply-headed to an interest driven economy is additionally provoking organizations to reconsider how they can fulfill request by making tweaked reactions to customer needs or assembling items with new and improved usefulness.

17.17. GETTING TO LITTLE AND MEDIUM-SIZED VENTURES

Global partnerships have the staff and the income to explore and receive the rewards of eco-productivity. In any case, numerous little and medium-sized endeavors (SMEs) feel they don't have time or cash for anything besides survival. By and by SMEs can likewise profit by eco-effective exercises. In Latin America, as exchange hindrances fall, organizations must be substantially more aggressive to endure. This is especially valid for the SMEs, which involve 90% of mechanical organizations. These organizations produce not exactly 50% of all mechanical yield, and an excessively high measure of contamination since they are under-directed and do not have the high perceivability of greater organizations.

However Latin American firms contend in a worldwide setting where eco-efficiency is fundamental to driving firms' business methodologies. In any case, they likewise work inside an unmistakably Latin American setting characterized by squeezing social needs, major natural and general medical issues, guidelines and constrained administrative implementation, macroeconomic approaches that support earth antagonistic practices, and restricted ecological infrastructure. Where cost-reserve funds don't put forth the defense, at that point SME interest in eco-efficiency relies upon the quality of government organizations and market request.

Regularly, the last is non-existent. Ventures in the area come up short on the neighborhood interest for, data about, or wellsprings of account for, cleaner items. Nonetheless, these difficulties can make openings: to profit by other nations' encounters, to jump frog innovation, to evade exorbitant missteps, and to build up a way to deal with ecological administration that assesses specific social needs and capacities.

At present, there is little market for eco-efficiency and governments are understanding that they should make that showcase. For example, the expense of getting to be confirmed by an authorize confirmation specialist, for example, ISO is restrictive for generally SMEs. Much of the time, the boundary is all the more an administration than a specialized issue. Ecological issues will in general be created by an absence of good administration in all regions, so such issues can't be fixed by a specialized change anywhere. Some enormous organizations are working with littler ones so all can profit by eco-effectiveness. Fortifying government foundations and business affiliations, making guidelines, making the confirmation procedure and ecological administration frameworks all the more generally accessible, instruction and preparing, and more exhibition tasks will help make the market.

Eco-efficiency arrangements must be adjusted to nearby conditions, both socially and financially. This implies taking a shot at best practices and great housekeeping regularly yields more noteworthy accomplishment than a huge innovation move or the usage of an Environmental Management System (EMS). The "familiarity" of numerous SMEs is another noteworthy boundary to encourage selection of eco-efficiency – frequently they don't cover government expenses or get water or vitality legitimately which gives them a lower cost structure than a contender.

Various advances have been distinguished for improving SME eco-effectiveness:

- SMEs can work with supply chains, neighbor organizations, or NGOs to make natural improvement and produce data that permit SMEs to show enhancements to the market (e.g., ISO 14001).
- Industry and government must produce home nation request by encouraging ecological advancements that foresee market requests and necessities.
- Engage the network and laborers. In OECD nations, the accessibility of open data concerning firm ecological execution, combined with nearby network activism, has been a noteworthy driver of improved execution.
- Financing organizations and safety net providers can require improved ecological execution and make help accessible.

A Mexican SME found a noteworthy wellspring of investment funds in diminishing its standardized savings installments when it was renamed as a non-dangerous work environment. Additionally, organizations will be bound to improve execution in the event that they can meet reasonable guidelines and are freely perceived for their endeavors.

The fundamental framework is another superseding concern. In the event that organizations are to isolate and oversee risky waste, for instance, there must be a spot to discard, reuse, or treat the waste. Extra strides making a course for eco-proficiency for SMEs incorporate all the more preparing and help, more prominent access to data, and the advancement of ease, powerful confirmation frameworks.

17.18. LIMITATIONS

Eco-proficiency has its commentators however this analysis frequently comes to fruition because of an absence of comprehension. William McDonough and Michael Braungart, pros in engineering and plan science, wrote in The Atlantic Monthly in 1998 that the idea was "lethally constrained." They asserted that eco-efficiency "guides us to confine industry and shorten development – to attempt to restrict the inventiveness and productiveness of mankind."

Truth be told – and this surprises a few earthy people – eco-efficiency sets no restrictions to development and no innate confinements on industry. It looks for increasingly proficient development, and in doing as such requires a lot of "innovativeness and productiveness" with respect to mankind. The creators' just proposed answer for the apparent issues with eco-efficiency was to change the term to eco-effectiveness!1 Eco-proficiency is likewise censured for not accomplishing every one of the objectives of practical advancement, as it needs expressed breaking points and it does not have a social side. In any case, it was never proposed to wrap up the entire idea for business; it was intended to give business an efficient path into reasonable advancement. It is a useful device, which enables organizations to improve business while having their impact in the natural side of maintainable advancement.

Numerous organizations that have grasped eco-efficiency have additionally proceeded to investigate the social parts of feasible improvement. Without anyone else's input, eco-efficiency can't adapt to business' destructive impacts upon nature. Monetary development is required to make occupations and open doors for developing populaces and to help individuals haul themselves out of destitution. One can envision a world where organizations are expanding their eco-efficiency by 5% per year while mechanical yield is expanding by 10% every year – and the planet isn't turning into a cleaner spot to live. Eco-efficiency isn't intended to be a comprehensive panacea. To be sure, while eco-efficiency is a significant advance in moving towards increasingly manageable examples of creation and utilization, it is essential to recollect that eco-efficiency is:

- not an accept the only choice available methodology; it should be coordinated in the organization frameworks.
- not an answer for every one of the deterrents to maintainability, nor a confirmation against disappointment; it is only one piece of the riddle.
- not an inflexible structure or a solitary technique, yet a strategy to improve rehearses in similarity with other corporate methodologies.
- not an authentic standard or an announcing design, however it can contribute towards improving and imparting results.
- not a cook-book with plans and arrangements one can pay off the rack; it needs organization explicit understanding and usage.

17.19. FOLLOWING STAGES

Ten years after the World Summit on Sustainable Development in Rio, obviously eco-proficiency works, building an incentive for clients and partners. The test presently is to move away from a consistence focused, emergency evasion mindset to seeing great natural and social execution as the fundamental establishment for the market worth and open notoriety.

A more grounded contention for eco-efficiency should be made to support associations, whose individual ecological effects are regularly moderately minor, however represent an enormous extent of monetary movement in created economies. Venturing up the pace will mean reinforcing the connection between eco-efficiency and investor worth and formulating new bookkeeping strategies to empower associations to survey eco-proficiency such that sounds good to the money related network. Organizations that are driving the best approach to eco efficiency perceive business' obligation to give authority in grasping eco-efficiency challenges.

Business is particularly set to comprehend client needs and how these can be conveyed through significantly more eco-effective administrations and items that make step enhancements in ecological execution. Business is likewise in a situation to make the center advances that are basic to reasonable improvement and to move thoughts and innovations around the globe. At last, business can utilize its showcasing abilities to illuminate customers who may once in a while be uninformed of the desperation and prerequisites of maintainable generation and utilization.

Eco-efficiency is likewise part of a more extensive idea – feasible utilization and generation. A comprehensive way to deal with the commercial center treats free market activity as one framework as opposed to separate, autonomous elements. In that capacity, the market can be a noteworthy power in improving supportability. Business must keep seeking after eco-efficiency and development, yet should likewise go past this to proactively set up associations with governments and non-legislative associations to carry items and administrations to meet existing and rising buyer and societal needs.

Business can likewise add to improving business sector structure conditions, favoring stable, defilement free financial ones that encourage positive change. They can give and advise customer decision, set up the value of the Earth's assets, and plan to make the market work for all. These standards are laid out in the WBCSD's production Sustainability Through the Market, distributed in 2002.

17.20. ECO-INNOVATION

"Human rights and the environment can be defended in a very direct way through the individual conduct of each and every business."

—Kofi Annan, Secretary-General of the United Nations

The center has moved from the operational parts of eco-efficiency toward eco-advancement and plan for condition. This has opened up much greater open doors for organizations to propel their eco-efficiency execution. For some organizations, the principle consequences for the earth really happened outside their fence-lines – either upstream in the crude material age and provider preparing stages, or downstream in the item use or transfer stages. The OECD distinguished advancement as the key driver for improving eco-proficiency and said that it was best invigorated by solid challenge, high factor costs, administrative motivating forces, a viable procedure of scattering "best practice," and the nearness of a decent atmosphere for development. What is required currently are radical enhancements, better item plan and further reception of eco-productive standards.

This is a valuable development action in helping individuals to think contrastingly and to target work cross-practically. The method is called fundamental innovative reasoning. You start by posting the fundamental components of an item, the two its physical segments and its properties, for example, shading and anticipated helpful life. You additionally take a gander at the item's prompt condition, again distinguishing the two its physical segments and its traits. At that point you control these components to think of something new, tailing at least one of four nonexclusive advancement designs.

17.21. THE FOUR NONEXCLUSIVE DEVELOPMENT DESIGNS

In growing new items, individuals instinctively will in general add highlights to a current item. While it is a coherent methodology, items may wind up convoluted and complex. Rather, attempt to evacuate a part or characteristic, even those that appear to be either unpalatable or essential. For instance, taking out the lead in petroleum or the sugar in soda pops. In dialog with your accomplice, plan something for your item to demolish it. Either a structure blemish, material imperfection or use defect – simply ruin it. For instance, scratch the CD, make the champagne level or take the hand off the watch. Re-imagine or invert develop the defective item, improving its ecological effect.

Augmentation includes making at least one duplicates of a current item segment and modifying those duplicates here and there to improve item quality. A great model is the "Gillette Double-Bladed Razor." The additional edge is an augmentation since it speaks to an additional sharp edge however it is set at an alternate edge. By what method can the increase procedure help animate new, earth inviting items?

By isolating a current item into its segment parts, you might most likely observe something that was a coordinated entire in a totally extraordinary light. You may then have the option to reconfigure those parts in unexpected ways. For instance, sound systems are frequently sold in individual parts to consider customization as opposed to in an incorporated framework. Think about an item and how its segment parts can help with new procedures or make side-effect cooperative energies.

You can regularly acknowledge noteworthy item advancement by relegating another undertaking to a current component of the item or its condition, in this way bringing together two assignments in a solitary part. A case of this methodology includes bringing together the defrosting fiber with radio gathering to dispose of the different radio reception apparatus. What about achieving new markets?

17.22. ADVANCEMENT MODEL 1

Sony: Innovations for the item use stage Sony is adopting a comprehensive strategy to improving items and administrations to limit ecological effects. Developments are improving creation, yet additionally diminishing natural effects in the items' utilization stage. This has prompted the improvement of various innovations that help diminish the natural effect ascribed to buyer conduct during item use, administration and transfer. Life-cycle evaluation calls attention to that the major natural weight of electronic items is because of the vitality devoured in the utilization period of the item life. European home customer hardware devour 36 TWh (terawatt-hours) and are anticipated to develop to 62 TWh by 2010. Sony's SDM-N50 fluid precious stone presentation fuses an interesting arrangement of client includes that straightforwardly lessens the vitality utilization of the item. The 12 mm, 3 kg presentation includes a vitality sparing infrared "client sensor" and a surrounding light sensor. The client sensor naturally changes the screen to rest mode (of a rate of under 3 watts) if nobody is sitting before the screen. The light sensor modifies the splendor (and along these lines vitality utilization) of the presentation as indicated by the characteristic surrounding light states of the room. Sony is likewise creating advances and administrations that add to the expansion of item lifetime. For instance, the ProGlobe TV is an European model plan gone for building up advances for improved workableness. The model advancements being created incorporate an extraordinary frame plan that permits administration designers to get to the interior center of the TV at the purchaser's home (conceivably decreasing the movement weight and cost of TV administration) and an oneclick, "eco-mode" work giving the client the alternative of diminishing the operational vitality utilization of the TV by 25 watts (by modifying the image brilliance). Likewise, the ProGlobe's new power circuit format gives an amazingly low degree of 0.5 watt remain by power utilization.

Scaling down and the move to advanced advances have made item fix an undeniably troublesome and particular employment. In Europe, Sony has presented the "trade and revamping program," a creative administration idea for the proficient fix of items with huge deals volumes, for example, the Walkman, Discman, PlayStation and cell phones. On account of an imperfection, a client can trade their item at the seller for a revamped model, or notwithstanding for another one (if a deformity is found inside the three-month ensure period from buy). Deformity items are gathered and revamped, broken parts are fixed or traded, and lodging parts are checked at a focal European area. Since industrial facility like fix is progressively effective, normal expenses have been decreased by 25%. Along these lines, more item parts and segments find the opportunity of a subsequent life. For instance, Sony is accomplishing a 68% reuse rate for the plastic parts of returned PlayStations. For all intents and purposes the majority of the rest of the parts that can't be reused are reused. www.sony-europe.com/head/environment.html

17.23. ADVANCEMENT MODEL 2

Dow: Innovation in administration rather than item Due to their cleaning productivity, chlorinated solvents are utilized in an assortment of ventures, including pharmaceuticals, plastics, dry-cleaning, and metalworking. They are likewise utilized for the creation of ordinary things, for example, the Swiss Army Knife. Now and again, particularly innovative applications, chlorinated solvents are essential. Be that as it may, in the mid-1980s new data about the potential perils related with the utilization of these solvents ended up accessible. In this manner administrative experts had some troublesome choices to make. A few nations decided to intensely confine the utilization of these solvents. Perpetually, this prompted the improvement of substitutes, yet huge numbers of these substitutes were untested items that were not really more secure. Nonetheless, the three fundamental chlorinated solvents that were available had been exposed to broad wellbeing studies and toxicological audits. Thus, their wellbeing profiles were notable, and successful security methodology could be embraced. Germany decided not to boycott the utilization of chlorinated solvents, but instead to necessitate that their utilization occur in a shut circle framework. Persuaded of the focal points offered by these items, The Dow Chemical Company stayed in the chlorinated solvents business. The organization was likewise sure that dealing with the dangers would give a superior, more secure choice than going to deficiently tried options with obscure dangers – following the standards of "Dependable Care." Dow started investigating new plans of action that would furnish clients with an ecologically and financially solid item in a sheltered and dependable way. In 1992, Dow Deutschland Inc. what's more, RCN (Recycling Chemie Niederrhein) established a joint dare to offer chlorinated solvents in a shut circle conveyance framework joined by help administration. The new framework offered not just the protected conveyance and reclaim of solvents yet additionally help to clients in the utilization of these solvents in shut circle gear. The organization was called SAFECHEM Umweltservice GmbH and turned into a completely claimed backup of Dow Deutschland Inc. in 1998.

Today SAFECHEM markets the SAFE-TAINER System crosswise over Europe. The framework focuses on two uncommonly planned twofold cleaned holders (one for crisp dissolvable and one for utilized) and every ha a coordinated drum for the dissolvable. This is planned to give security against unplanned harm, release, and spillage. The offering, be that as it may, involves unmistakably something beyond safe holders; the SAFETAINER System is a totally new plan of action whose achievement is established on two contemplations: environmental and monetary manageability. With the usage of shut circle frameworks, dissolvable utilization has dropped impressively, because of the decrease in discharges. The dissolvable stays for expanded periods (a year and that's just the beginning) in the shut cleaning hardware. The objective of this methodology is to utilize the dissolvable for whatever length of time that conceivable in a reusing circle at the client's site before outside reusing. SAFECHEM is an administration-focused business, selling least solvents and most extreme help administrations to enable the clients to accomplish objectives. SAFECHEM and Dow offer this administration stage in Germany, Switzerland, Austria, Italy, Spain, France, the UK, Sweden, Benelux and Portugal. It is on account of the administrative improvement supporting the utilization of chlorinated solvents in shut circles, instead of their forbiddance, that the possibility of SAFECHEM as administration and arrangement supplier has demonstrated such a huge achievement in Europe (www.dow.com/safeche).

17.24. ADVANCEMENT MODEL 3

Dow: Innovation underway and crude materials Dow declared as of late that it would start making elective items that will go up against customary wood-based items. The organization obtained a lion's share of the benefits of Isobord Enterprises, Inc., and will create built composite boards produced using waste wheat straw and Dow's polyurethane tar. The item will be advertised under the name WOODSTALK and can be utilized for kitchen counters, racking, prepared to-amass furniture, cupboards, entryway centers and floor underlay. "These composite boards give an ecological advantage to buyers since they are produced using a yearly sustainable asset and have the exhibition attributes of customary wood-based boards. Utilizing straw for such boards decreases the need to consume it in the fields, which makes air contamination. This item fits well with Dow's responsibility to reasonable advancement," as per Brad Money, general director and leader of Dow BioProducts. Includes Money: "Purchasers progressively are searching for items that diminish the ecological effect. Woodstalk items do only that by using the every year inexhaustible lingering wheat straw. Woodstalk has conceivably a solid business offer for shoppers and contractual workers." Through numerous outlets in North America, home-improvement retailer Home Depot appropriates the Woodstalk items. Says Ron Jarvis, Home Depot's Lumber VP of Merchandising: "We trust Dow's assets, know-how and notoriety will be a solid factor in bringing earth monitoring wood substitution boards more into the standard with buyers." By purchasing a current, present day fabricating office whose workers have the skill and mastery in creating these boards, the organization abstained from "working without any preparation in a green field circumstance," says Money. Dow's skill in procedure innovation and polyurethane covers joined with that of the current workforce in composite board generation makes a perfect domain for improving and building up the business. The confirmation is in the framing. These items have gotten a few natural honors, including the Sustainable Development Award of Excellence from the Manitoba Round Table and the Salute to Clean Air Industry grant from the Manitoba Lung Association. Sparing trees – and lungs – is a triumphant mix. Dow BioProducts Ltd., a completely claimed auxiliary of Dow Chemical Canada Inc., www.dow.com/dow_news/prodbus/2001/20010914a_pb.html

17.25. DEVELOPMENT MODEL 4

Lodging configuration: Using an open door A creator in the Netherlands who needed to spare space in a lodging venture joined a sink and can at the hip and discovered a basic structure that could spare a large number of gallons of water. Not terrible when you think about that, constantly 2020, 66% of the world's kin will live in a water-starved region. For one structure engineer, the appropriate response came in utilizing one procedure's loss to control a related yet at the same time altogether different procedure. The creator had made a mutual structure for a restroom sink and a can, which was initially planned as a space-sparing element. In taking a gander at what he had made, he saw the chance to lessen the measure of family water required. Some 30% of family unit water utilized is truly flushed down the latrine. With a couple of changes to the structure, the water from the sink was gathered in a repository and therefore utilized in the latrine instead of the new perfect water. This reuse of purported "dark water" spared critical measures of water and decreased the measure of water that entered sewage treatment plants.

17.26. ADVANCEMENT MODEL 5

SoftAir™: structure for less assets SoftAir™ reexamines furniture plan – another strategy to use an endless, 100% eco-accommodating and totally free characteristic asset: air! The SoftAir™ furniture depends on a delicate, earth adjusted, innovative plastic material and another assembling method. Together they structure the fundament of SoftAir Technology (pat. pend. int.). One approach to improve our prospects is to fabricate just items that require couple of assets, are sustainable, biodegradable and can be destroyed and arranged for reusing. To make a SoftAir™ couch you need just about 15% of the materials, vitality and generation assets required to make a conventional couch. The investment funds become much more noteworthy when you take a gander at transportation and capacity. Everything means a decrease of 85% of the aggregate sum of vitality devoured by a couch during its lifetime. The delicate plastic at the core of the SoftAir™ furniture is probably the most perfect plastic there is. Polyolefin is 100% recyclable. In case of flame it creates no poisonous exhaust. It consumes in a moderate and quelled style and produces little smoke. The whole gathering is painstakingly tried both concerning quality and capacity. Reusing is regularly an unpredictable and expensive procedure. For SoftAir™, be that as it may, this expense is immaterial (www.softair-furniture.com).

17.27. DEVELOPMENT MODEL 6

Portage and DuPont: Creative participation A captivating model is a course of action by Ford and DuPont in the European market. Generally, DuPont was glad to sell Ford all the paint it could regarding Ford's vehicle painting tasks. In a novel course of action, Ford contracted with DuPont to go to its plant and legitimately paint the vehicles. This adjusted the requirements of both the client and the provider – to use as meager paint as could be allowed and along these lines lessen the effect of the item and the procedure on the earth. Agreeable plans like this exhibit the innovativeness characteristic in the requirement for eco-development (www.greenmarketing.com/gemipreconference.pdf).

17.28. CONTEXTUAL ANALYSES AND MODELS

17.28.1. BASF and an Eco-Efficiency Apparatus

BASF has built up a vital life-cycle apparatus for looking at the financial and natural focal points and hindrances of items and procedures. The device utilizes a two-hub chart that plots the two estimations of natural weight and absolute expense. When evaluating natural effects, six classifications are analyzed: crude material utilization, vitality utilization, air and water emanations and transfer, potential lethality, potential dangers and land use. The aftereffects of eco-efficiency examination are helping the organization to improve items and procedures by showing precisely where improvement would profit the earth most and what its monetary results would be. The framework likewise permits BASF to characterize and screen innovative work targets (www.basf.com).

17.28.2. Cemento de El Salvador Heaters and Woodlands

Cemento de El Salvador S.A. de C.V. (CESSA), has been utilizing an eco-efficiency activity to spare vitality and improve natural effects. The focal point of this exertion has been the establishment of another creation line dependent on dry heaters. Already, three of CESSA's heaters had been of the wet revolving type, extremely wasteful in fuel use. By utilizing this new innovation, introduced in 1997, CESSA has diminished power utilization, CO_2 emanations, and nitrous oxide particulates since dry heaters utilize less vitality. CESSA is additionally reforesting a normal of four hectares for every year, and every hectare of woods counterbalances four tons of CO_2 outflows. A considerable lot of the reforested zones were previous mining pits, deserted streets, or basically handles that had been deforested (www.cessa.com.sv).

17.28.3. Chaparral Steel Company Recycling Squanders from Steel and Bond

Chaparral Steel Company started its "Frameworks and Technology for

Advanced Recycling" (STAR) program mutually with TXI Cement. The framework aims to process and preserve steel, producing the minimum amount of contamination and waste materials. As a component of the STAR venture, Chapparal and TXI Cement protected "CemStar," a procedure that includes slag, or stony waste issue isolated from metals during purifying or refining of metal, to the crude material blend, yielding bigger clusters of brilliant Portland bond without bargaining its attributes. This side effect to asset transformation monitors common assets while lessening vitality prerequisites by at any rate 10%, decreasing CO_2 and NOx (oxides of nitrogen) outflows. Moreover, the estimation of the slag has ascended by a factor of 20 in respect to its incentive as a street development material, its past use, and TXI accepts most bond furnaces can expand their ability by 5% to 15% with a little capital venture and low steady expenses (www.chaparralsteel.com www.deh.gov.au/industry/index.html).

17.28.4. CH2MHILL and Nike Eco-Proficiency Through Inventory Network Measurements

Nike's enormous and differing inventory network sellers are a piece of its supportability activity which expects to protect and save water and diminish the ecological effects of wastewater release underway tasks. CH2MHILL has worked with more than 250 of these merchants to build up a database, issue revealing structures to processing plants, gather and approve quarterly information from production lines, and create quarterly corporate ecological reports. Under the program, CH2MHILL set up a gauge for strong waste delivered in 45 footwear-processing plants in Asia. The objective is to diminish strong waste creation and increment reusing, while at the same time recording and covering progress (www.ch2m.com; www.nike.com).

17.28.5. Dow Europe Eco-Compass

Dow Europe has created "Eco-Compass," a device utilized side-effect and procedure creators to describe new items against previous renditions of the results of their rivals. A position of 2 implies it is the equivalent, 0 is a factor of two more awful, 4 a factor of two better, and 5 a factor of 3 better (www.dow.com).

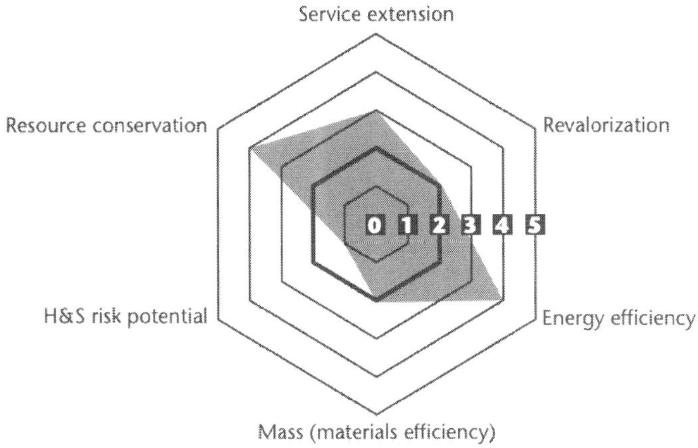

17.28.6. DuPont: Reducing Harmful Outflows

A DuPont agrarian items group from LaPorte, Texas, diminished its poisonous emanations by 99% through shut circle reusing, off-site recovery, selling previous squanders as items, and crude material substitution. Generally speaking reserve funds included $2.5 million in capital and more than $3 million in yearly working expenses (www.dupont.com).

17.28.7. DuPont Focused on Ecological Objectives

An attention on focused natural and vitality objectives has brought about a 91% decrease in worldwide air cancer-causing agents, a 47% decrease in worldwide unsafe waste, level vitality use for as long as decade while generation expanded over 30%, and a 68% decrease from 1990 in CO_2-proportionate ozone harming substance outflows from its worldwide assembling activities. While DuPont has obviously made significant speculations to accomplish these outcomes, it has likewise decreased its remediation and end-of-the-pipe squander treatment costs by over $250 million every year contrasted with 10 years back. DuPont has spared over $1.5 billion in vitality expenses contrasted with where the organization would be today if vitality utilize expanded directly with creation. DuPont has additionally created over $40 million in genuinely close term incomes from the clearance of ozone harming substance emanation decrease credits (Chad Holliday before the TERI Delhi Sustainable Development Summit).

17.28.8. Fundación Entorno Eco-Efficiency Unit

In 1999, the Spanish accomplice association Fundación Entorno propelled a pilot venture with 12 Spanish organizations to show the reasonability of items, procedures, and administrations created through eco-efficiency conceptualization. Because of this work, the Eco-Efficiency Toolkit 1.0 programming was created to help organizations improve and measure their eco-productivity. The three modules take a gander at eco-plan, benchmark and eco-effectiveness. The Eco-Efficiency Toolkit 1.0 has been sent to in excess of 800 organizations. The results include:

- The electrical building and hardware organization SIEMENS, AG utilized this toolbox for its Barcelona industrial facility, which makes train, tram, and streetcar motors. Siemens directed an examination of procedures and items created in-house to figure out which of these required further enhancements, focusing especially on procedures that utilized a great deal of vitality. The examination helped the organization select elective advancements that have streamlined development, cut vitality utilization by 45%, broadened the working existences of key hardware and spared 8,803.70 euros ($7,847).
- One of Spain's biggest counseling gatherings, TYPSA Group, utilized the Eco-efficiency Toolkit 1.0 to dissect a water cleaning plant for a creating nation venture. The pack helped them to distinguish and present eco-proficiency upgrades in every movement all through the procedure and dole out natural expenses. Vitality utilization was improved by 66% and the expense of the last item improved by 85% – showing the advantages of eco-efficiency in creating nations (www.fundacion-entorno.org).

17.29. HEWLETT-PACKARD USING REUSED HP ITEMS

In 2003, HP met its objective to present an equipment item containing reused material substance recouped from end-of-life HP items. Designers from numerous product offerings collaborated with HP's Planet Partners™ reusing group to create and qualify a plastics definition that could supplant virgin plastic with material acquired from HP end-of-life items. The group built up a material called reused polyethylene terephthalate (RPET). RPET is a mixed material that contains plastic from HP print cartridges reused in HP's Planet Partners™ program and post-buyer reused plastic drinking bottles. RPET is utilized at full generation levels in the carriage spread for two HP scanners, the ScanJet 4500 and 5550. This decreases virgin materials use, and with extra volumes and experience, it could prompt cost reserve funds. Two extra scanners utilizing the reused material were propelled in 2004. One of these models is relied upon to use around 100 tons of RPET during the following two years.

17.30. HITEGA AND ECO-PROFICIENCY

Hitega (Hilados y Tiejidos Garib), a Chilean material plant creating colored yarn and texture made of polyester and rayon, utilized eco-efficiency to recognize enhancements in the utilization of water and vitality, and the measure of suspended solids in the emanating stream. A few outcomes included reusing the relaxed water used to cool the color shower; reusing the water supply for forced air systems in the turning and weaving rooms; and improving conditioner recovery and administration. It additionally introduced screens in color room channels to lessen the measure of suspended solids in emanating (fluid waste or sewage released into a waterway or the ocean), and redid its upkeep intend to address spilling steam traps. The subsequent ecological advantages of these activities were water, vitality and compound protection, and diminished outflows and profluent borne solids. The vast majority of the measures received had restitution times of two years or less. The reusing of color cooling water, for instance, cost $750, however, conveyed investment funds of $400 per year (www.hitega.cl; www.bsdglobal.com/viewcasestudy.asp?id=71).

17.31. MINETTI GROUP USING WASTE AS FUEL IN CONCRETE FURNACES

The Argentine organization Ecoblend, some portion of the Minetti Group of concrete organizations, has built up an innovation to utilize mechanical waste as a fuel in bond furnaces. This elective fuel procedure has a few advantages. It rations non-renewable energy sources, for example, gas or diesel. It spares organizations the errand of discarding their very own portion squanders, and it expels squanders from nature, with no remaining material, for example, slag, which is frequently the situation with different advancements. The procedure additionally adds to the decrease of carbon dioxide, sulfur and nitrogen oxide discharges. The utilization of these waste items in concrete ovens keeps away from discharges that would be created through cremation. Since 1997, the utilization of mechanical deny as an elective fuel in bond ovens in Argentina has spared 49,350 m^3 of non-renewable energy sources and an aggregate of 122,500 tons of modern waste from landfill (www.ecoblend.com.ar/viewpg.asp?Cod=420).

17.32. NORSK HYDRO ALUMINUM REUSING

Norsk Hydro embraced a real existence cycle investigation way to deal with aluminum creation, assessing generation and potential eco-proficiency openings from a worth chain and frameworks point of view. Specifically, life-cycle investigation uncovered that the reusing of aluminum sets aside to 95% of the vitality utilized for essential generation. Therefore, the organization started to build up its reusing and remelting idea for aluminum. During the 1990s, the organization significantly expanded its remelting ability to build the measure of reused aluminum gave. In 1992, remelted aluminum represented 180,000 tons, while essential aluminum levels were a lot higher, at 610,000 tons. By 2001, essential levels had developed by 29% because of the expansion popular to 800,000 tons. In any case, the reused and remelted substance developed by over 400% in a similar period to 750,000 tons (www.hydro.com/en).

17.33. PHILIPS LIGHTING'S MINIMIZED GLARING LIGHT INNOVATION

Philips Lighting imagined an item that is offering impressive vitality and cost sparing potential. Conservative bright light innovation offers lights that last multiple times longer and utilize 75% less vitality than brilliant bulbs. This innovation can possibly make impressive contrasts in vitality utilization – and costs. For instance, it is anticipated that New York's Times Square could spare 23 million kWh and $5.8 million every year with the bulbs and the Strip in Las Vegas could spare 105 million kWh and $8.4 million yearly (www.lighting.philips.com).

17.34. SC JOHNSON'S GREENLIST FOR PICKING FIXINGS

SC Johnson's Greenlist procedure enables settle on to better ecological decisions as they define items. The Greenlist™ is the main Environmental

Initiative in SC Johnson's Five Year Strategic Environmental Plan. Following the aftereffects of its Greenlist™ procedure for quite a while, SC Johnson has various great stories. The accompanying speak to accomplishment in two of its real brands. Through the Greenlist™ procedure, SC Johnson built up another recipe for its Windex® window more clean. The new recipe utilizes crude materials that are better for the earth and surpasses shopper desires by not streaking and being anything but difficult to wipe. The objective for reformulating the item was to improve cleaning execution while diminishing unpredictable natural mixes (VOCs). In 2002, 181,800 kg of VOCs were expelled from the equation while keeping up the "sans streak" recipe. In 2004, another 818,000 kg of VOCs were expelled and cleaning execution was improved by 30%. This was finished utilizing a patentable mix of surfactants, solvents low in VOC content and a one of a kind polymer. Simultaneously SC Johnson evacuated an unstable natural dissolvable called 2-butoxy ethanol that was as of late delegated "poisonous" under the Canadian Environmental Protection Act. Since the new equation was presented in September 2005, Windex® deals have expanded by 8% and piece of the overall industry has expanded 3.9 focuses.

17.35. SC JOHNSON: REDUCING WASTE, HAZARD, AND COST

During the 1980s and 1990s, the organization focused on eco-efficiency by lessening waste, hazard and cost, and by making the executives frameworks to quantify decreases. Somewhere in the range of 1990 and 1995, generation expanded by half, while squander emanations were cut considerably. In the previous 10 years alone, the organization has cut squanders from items and procedures bringing about yearly cost investment funds of more than $125 million (www.scjohnson.com).

17.36. SONAE: SHOPPING CENTER ADVANCEMENT

Sonae separates itself from other Portuguese shopping center contenders through advancement. Using more sunlight and low-vitality light fittings, upgraded fabricating the board frameworks, a modest and inventive cooling framework, and flammable gas and co-age frameworks, Sonae is setting aside cash and the earth. Furthermore, 50–75% of the wood utilized in its wood board business comprises of wood squanders and side-effects. At one strip mall it has planned imaginative waste gathering, dealing with and reusing plans that happen underground, far out. At a similar site, cooling is accomplished by ceaseless watering of a 110-meter-long glass bay window. This brings down encompassing temperature while making a group satisfying special visualization as daylight backdrop illuminations the water streaming over the glass (www.sonae.pt).

17.37. SONY'S NEW FIX PLOT

Scaling down and the move to advanced advances have made item fix an inexorably troublesome and particular occupation. In Europe, Sony has presented a trade and revamping program for the fix of items with enormous deals volumes. Since the origin of this program, costs have been diminished by 25% (www.sony.net/index.html).

17.38. TAIHEIYO CEMENT CORPORATION: ECO-CONCRETE

The organization has built up an eco-concrete procedure that utilizations squander from society and different enterprises, and in which dioxins and unsafe substantial metals are caught. Through the assembling of eco-bond, the organization adds to both vitality preservation and CO_2 decreases (www.taiheiyo-cement.co.jp).

17.39. TEPCO: ECO-PRODUCTIVE VITALITY

The Tokyo Electric Power Company (TEPCO) has actualized different CO_2 discharge decrease estimates both on the free market activity sides. On the supply side the organization has been attempting to improve the effectiveness of warm power age and to decrease transmission and circulation misfortune rates. These upgrades, combined with hydroelectric generation, lessen fuel costs and the utilization of significant assets. These endeavors additionally decreased CO_2 discharges in 2000 by what could be compared to about 14.6 million tons contrasted with 1990. On the interest side, TEPCO advances and supports the vitality preservation endeavors of its clients to upgrade in general vitality productivity. In October 2000, in light of reviews of its customers' eagerness to pay for sustainable power source, TEPCO propelled its "Green Power Fund" to further advance sustainable power source (www.tepco.co.jp).

17.40. UNILEVER: PROJECT MEDUSA TO SPARE WATER

Unilever has many assembling tasks in Latin America – a district encountering issues of water accessibility. As water is a fundamental utility or fixing to Unilever's procedures, the organization tested its sustenances assembling locales to diminish their water engrave by 1,000,000 m_3. Task Medusa was propelled in April 2003. The task included preparing site designers and wellbeing/wellbeing/condition organizers at the 17 sustenances assembling destinations on eco-efficiency just as at a two-day workshop. The designers and facilitators contributed instances of good water advancement rehearses and shared their encounters. Locales at that point chosen the great practices that connected to their tasks and assessed their normal water investment funds – the site's "Medusa contract and activity plan" for water reserve funds. Observing, site water adjusts and extra metering at each site are additionally key pieces of task Medusa. Unilever's biggest assembling site in the Goiânia locale in Brazil has announced water reserve funds of more than 500,000 m3 because of the usage of its Medusa activity plan. Altogether the flow gauge of water investment funds in general in Latin America is moving toward the objective of 1,000,000 m3.

17.41. UTC: TEAMWORK TO HAIL PROFITABLE ENHANCEMENTS

One of United Technologies Corporation's locales wiped out just about 40,000 gallons for every time of waste water and spared over $50,000 every year with an essential change in the manner its deals with its test cells, underground stockpiling tank and waste streams. To accomplish this, the site now just uses its underground tank for crisis purposes – and not for customary waste or water, as it was doing. It additionally gathers squander liquids (powers and oils) independently for re-use, reusing or recovery. Given these changes, the site foresees it will take out an extra 75,000 gallons of waste for every year and spare $150,000 every year. These eco-efficiency upgrades were recognized during one of UTC's Kaizen occasions. UTC conducts condition, security and wellbeing Kaizen occasions over its destinations to drive consistent improvement in its ecological, wellbeing and wellbeing execution. The occasion centers consideration around explicit procedures and the Kaizen group comprises of all gatherings keen on that procedure, from the board to administrators. The group directs an organized, centered assessment of an assembling or administration process, coordinating Kaizen systems and adaptable assembling critical thinking methods with EH&S hazard appraisal. The reason for the occasions is to distinguish and kill or diminish ecological, wellbeing and security risks.

17.42. THE WAREHOUSE'S ENERGY MANAGEMENT TEAM

New Zealand retailer The Warehouse's Energy Management Team has been sparing the organization a ton of cash. The group has structured a product program that consequently controls lighting, warming and cooling enabling the whole chain to be worked from one focal office. Shaded screen presentations give 3-D portrayals of the stores, their floor plans, and the design and status of the lighting, warming, ventilation, cooling and other vitality utilizing gear, giving an easy to understand interface to staff. Vitality utilization, estimated each half hour in each store, is accessible on a safe site kept running by a metered provider. The stores have no light switches – all is constrained by the framework, which reacts to the lighting levels and time of day. More than seven years the cooled volume of stores expanded significantly, however was not coordinated by a comparative ascent in power utilization. Truth be told, normal power utilization dropped from a unique 200 kWh/m^2 to 100 kWh/m^2 (www.thewarehouse.co.nz).

17.43. QUOTES AND SPEECHES

"Eco-efficiency catches at a glance the balance business strives toward: sound ecology and profitable operations."—Bjorn Stigson, President, WBCSD (Walking the Talk).

"Business cannot achieve eco-efficiency alone. Progress requires going beyond the internal actions of companies; it needs governments to create an enabling framework which allows individual companies and markets to become more eco-efficient."—Bjorn Stigson, President, WBCSD – The World Bank Group Environment Strategy, June 6, 2001, www.wbcsd.org/DocRoot/41GhZ9EHj28TqoLfGJpk/ee-bs-worldbank.pdf.

"The concept has moved from preventing pollution in manufacturing industries to becoming a driver for innovation and competitiveness in all types of companies. Eco-efficiency applies throughout a company. It is not limited to areas within a company's boundaries, such as in manufacturing and plant management. It is also valid for activities upstream and downstream and involves the supply and product chains."—Bjorn Stigson, President,

WBCSD – The World Bank Group Environment Strategy, June 6, 2001, www.wbcsd.org/DocRoot/41GhZ9EHj28TqoLfGJpk/ee-bs-worldbank.pdf.

"My colleagues and I are totally committed to a business strategy that generates profits while contributing to the wellbeing of the planet and its people. We see no alternative."—Mark Moody-Stuart, Chairman, Shell "Environment will represent one of the biggest possibilities for technical and leadership-based innovation – and profitable companies – which the world has ever witnessed."—Percy Barnevik, ABB, The World Bank Group Environment Strategy, www.wbcsd.org/DocRoot/41GhZ9EHj28TqoLfGJpk/ee-bs-worldbank.pdf.

"Just as the Stone Age did not end for a lack of stones, the Oil Age will not end for a lack of oil."—Preserving the Environment and Raising Living Standards Through an Open Food System. Incorporated to the 32nd International General Meeting Pacific Basin Economic Council. Ernest S. Micek, Chairman and CEO, Cargill, May 18, 1999, www.cargill.com/today/speeches/micek3.htm.

"We can support SMEs' adoption of eco-efficiency in two main ways. First is the reinforcement of the supply chain management. The larger companies can play a very important role in encouraging its SME suppliers to practice eco-efficiency. Second is the support by the government, including training, financing and incentives."—Yao Sheng Chen, Chairman, Taiwan Prosperity Chemical Corp. and Taiwan Cement Group.

"If it is better, and more profitable, to do things responsibly, people will do things responsibly. But without the right framework and structures in place, the motivation will be lacking. Because there is generally little pressure from authorities or civil society, many small or medium-sized businesses practice business in an informal way. They don't pay taxes; they obtain water or energy illegally; and they don't pay social security or minimum wages. And that way they can start out with a lower cost structure than a competitor who is determined to do things legally. This is why it is so important to set the right framework so that others don't get a free ride."—Maria Emilia Correa, vice President of Social and Environmental Responsibility, GrupoNueva (Walking the Talk).

"Eco-efficient innovations allow companies to use a range of inputs more productively – from raw materials to energy to labor – thus offsetting the costs of improving environmental impact. Ultimately, this enhanced resource productivity makes companies more competitive, not less."—Michael Porter.

"Bearing in mind the nature of our businesses, we founded the long-term performance strategy for Suez Lyonnaise des Eaux on eco-efficiency. This concept links competitiveness and value creation to protecting the world's equilibrium and enhancing our understanding of environmental challenges. Sustainable growth must thus be part of business strategy, integrated in all decision-making processes, and demonstrated through concrete action."—Thierry Chambolle, senior vice President, Environment and Technology, Suez Lyonnaise des Eaux.

"Our experience shows that sustainability does not require sacrifice or deprivation. Rather it means exploiting the latest technologies, reducing waste, reusing resources, adopting lean production and better logistics practices and making sound investments in efficient productive capacity. In fact, eco-efficiency and sustainability are not free; for a market leader they are actually more than free because they provide a real competitive advantage."—Pasquale Pistorio, President and CEO, ST Microelectronics.

"Our business challenge is to deliver competitive economic value, and at the same time operate in an environmentally sound and socially responsible manner. I strongly believe that excellence in environmental approach and solutions, in social responsibility and in economic performance are complementary, and not conflicting. In periods with substantial and rapid business restructuring, the need to provide practical proof of this complementarity is becoming increasingly important."—Egil Myklebust, President and CEO, Norsk Hydro.

CHAPTER 18

HR Driving Sustainable Business Practices

18.1. INTRODUCTION

There's a dire test for HR pioneers: ensuring their businesses expect and get ready for the inescapable impacts of sustainability super powers. As globalization, moving socioeconomics and rivalry for the world's exhausting assets constrain transformational change, organizations will require edified and sustainability sagacious administration to flourish in this overcome the new world. HR has a noteworthy part to play to adjustability to these rising substances.

Sustainability superpowers will influence the capacity of businesses to succeed and flourish over the coming decades. The worldwide population is relied upon to expand from seven to nine billion individuals by 2050, compelling organizations to reexamine themselves to secure access to assets and the social permit to work and develop.

Similarly as concerning, the salary imbalance hole keeps on augmenting, additionally undermining the social union. Versatile businesses that suspect, anticipate, and lighten these patterns will fabricate strength and upper hand while adding to a more reasonable society.

Yet, for businesses to prevail in this new worldwide setting, present, and future pioneers will require a large group of new abilities and capabilities – and hierarchical motivating forces should line up with key sustainability needs. HR experts can impact and empower this key move.

18.2. VALUES-BASED COMPANIES

Employees progressively like to work for businesses that mirror their qualities. Twenty to thirty-year-olds need to construct professions with businesses that are having a beneficial outcome. Mid-to-late vocation experts, then again, may mull over leaving their businesses to satisfy a want to add to more prominent's benefit.

So how might we develop HR's part inside businesses to draw in and hold these employees? Corporate social responsibility (CSR) is firmly related with high employee engagement and in this way hierarchical profitability and development. Organizations with solid CSR and sustainability programs appreciate higher assurance and faithfulness. HR experts possess the key work area to help encourage great inward CSR results and incorporate it with the worker offer.

18.3. SUSTAINABILITY CENTRAL TO HR PORTFOLIO

The business train of sustainability and corporate social duty has just truly developed in the course of recent decades. Initially, CSR was an altruistic, consistence or operational issue, yet given the new business setting and rising desires from financial specialists, clients, government and workers; CSR is progressively a key issue for organizations.

For the 21st-century association, social and natural patterns exhibit both worldwide dangers and openings, which influence operations and in addition providers and clients. Social and natural issues are currently, like never before, official and meeting room issues. Thusly, CSR contemplations are fundamental to the HR portfolio.

18.4. TOOLS TO HELP

There are numerous assets to enable HR administrators to end up plainly capable in installing CSR into the employee lifecycle and experience (see rundown of assets toward the finish of this article). On the off chance that done accurately, HR can bolster an association to end up noticeably future-fit, installing social and ecological elements into:

- Corporate reason, vision, mission, qualities, and technique;
- Employee set of principles;
- Workforce arranging and enrollment;
- Orientation, preparing and competency advancement;
- Compensation and execution administration;

- Change administration and corporate culture;
- Employee contribution and engagement;
- Employee interchanges;
- Celebrating achievement.

These means are inseparably connected to HR's center skills, including the capacity to build the engaging quality of the business to attractive potential employees; to make a workforce design by distinguishing future ability needs to help the association's objectives and targets; and to recognize authoritative learning needs lined up with the business technique.

18.5. CSR CRITICAL TO SUCCESSION PLANNING

Not far from this issue lies the matter of CEO succession planning. With CEO turnover at about 16 per cent and retirements a top driver of CEO renewal in 2014, according to outplacement consulting firm Challenger, Gray & Christmas – and given the relationship between effective sustainability performance and firm success boards and their HR advisors must ensure they have adequately included sustainability and CSR factors in their succession plans and search criteria. There are six critical CSR requirements for CEO succession planning and recruitment:

- "Values" role model;
- Externally aware;
- CSR strategist and change manager;
- Collaborates with stakeholders;
- Catalyst and advocate;
- Develops responsible leaders.

As CEOs set the tone at the top, getting this right will be essential for CSR and sustainability progress at any organization.

18.6. MOTIVATING SUSTAINABILITY-CONSCIOUS BEHAVIOR

HR managers will also need to play a lead role advising on sustainable pay metrics and bonus-able goals. Organizations have a long way to go getting the right incentives in place. While many companies include social and environmental factors in executive compensation plans, most are compliance-focused and backwards looking and few have actual targets.

HR managers can play a stronger role building organizational insights in this area and aligning incentive compensation with strategic sustainability objectives.

A value-added HR professional will support the organization to anticipate and manage these profound labor market and societal shifts to foster business and social success. HR managers must find ways to bring CSR and sustainability into scope when sourcing and optimizing talent. This helps achieve organizational outcomes to realize CSR's power as a top driver of employee engagement and retention.

18.7. EMBEDDING SUSTAINABILITY IN HR (TABLE 18.1)

Table 18.1: Sustainability Strategy Implementation in HR

Critical success factors for implementing sustainable business practices	**Role of HR**
Creation of a sustainable vision (core values, core purpose and envisioned future), mission and strategy	Integrate the sustainability theme with the vision of the organization Ensure alignment of employee behaviors
Creation of a sustainable vision and mission	Facilitate development of a sustainable vision, mission, and strategy Support creation of the sustainability strategy by bringing in employee perspective Communicate vision and mission across all levels in the organization

Building of an aligned leadership pipeline	Align the existing leadership to the sustainability agenda Build a leadership pipeline by institutionalizing leadership development systems
Creation of appropriate organization structure and governance mechanisms to drive the sustainability agenda	Facilitate creation of an appropriate organization structure to drive sustainability initiatives
Development of key HR capabilities to drive sustainability and create sustainable HR systems and processes	Understand sustainability in the context of the business Build sustainable HR systems and processes
Ability to attract, engage and retain the Right set of people and align them to the sustainability agenda	Manage the organization's expectations
Creation of an agile organization, which is able to learn and adapt to changes triggered by internal and external environment	Formalizing a process of informal learning and knowledge sharing managing personal journey of employees toward a sustainability centric mindset through effective change management
Engagement with external and internal stakeholders – employees, future employees, supply chain, customers, community at large, NGOs and pressure groups, regulators and suppliers of equity and debt capital; constitutes key to inclusive growth of the organization	Spearhead the stakeholder engagement process, coordinate among the stakeholder

Stakeholder communication – Communication of the organization's Sustainability Performance annually to stakeholder group is key element of the growth process; helps in brand building, banks, local community, and regulators as well as in benchmarking sustainability performance among peer groups	Lead the Sustainability Report development process after mapping the key material performance indicators, measurement and comparing these against own set targets

Source: Ernst & Young (2017).

CHAPTER 19

Critical Success Factors for Implementation of Sustainable Business Practices From a People Perspective

HR can enable businesses to prevail with regards to executing sustainable business practices. In spite of the fact that numerous businesses influence an actualize, to gauge and coordinate it (Figure 19.1).

Figure 19.1. Critical success factors while implementing sustainability practices.

Source: Ernst & Young (2017).

Our experience demonstrates that businesses need to address the accompanying basic achievement factors from a people point of view to drive reasonable business practices. HR assumes a key part in powerful administration of these basic achievement factors:

- **Creation of a Sustainable Vision and Mission:** To start with, an association can't actualize economical business practices on the off chance that it doesn't embrace them in its ethos depicts its dedication toward feasible business objectives.
- **Building of an Adjusted Initiative Pipeline:** Organizations must

adjust their authority to drive the sustainability motivation. It is the initiative, which is in charge of motivation. Pioneers are good examples for different workers and henceforth, assume a key part in setting the correct cases and heading for the association. Be that as it may, it is similarly imperative to make an administration pipeline adjusted to the sustainability logic to convey forward the association's vision and mission later on.

Making of a proper association structure and administration systems to drive the sustainability plan: Organizations must roll outfitting auxiliary improvements and make solid administration and coordination components to guarantee that reasonable responsibility is built up and sufficient power and assets are assigned to drive the sustainability motivation. The association needs to distinguish objectives and objectives, track advance, and report results to guarantee that the sustainability motivation is on track. Without a formal fruitful execution of reasonable business practices.

Improvement of key HR abilities to drive sustainability and make sustainable HR frameworks and procedures: While the part of HR is basic in driving feasible business practices at all levels and making reasonable HR frameworks and procedures, numerous HR experts do not have the information about sustainability, how it impacts business and how to drive authoritative change. As an outcome, HR regularly neglects to win the trust of business in spite of being the common accomplice for driving sustainability. Moreover, HR experts need to create useful capacities to make reasonable HR frameworks and procedures.

Capacity to draw in and spur workers and adjust them to the sustainability motivation: Apart from the authority, businesses can understand their sustainability plan just in the event that they pull in and hold the correct ability to guarantee acknowledgment of the association's objectives and targets and to take an interest and convey forward the motivation on a proceeded with premise.

Formation of a nimble association, which is capable to learn and adjust to changes activated by the interior and outside condition: The way to a feasible association additionally includes the capacity to persistently learn, adjust and oversee change. Businesses, which can share and recreate best practices and can persistently take in and enhance from past encounters are maintainable. Sustainability itself includes a change in the association's outlook. There is a need to make components for ceaseless learning and successful change administration inside the association.

19.1. CREATION OF A SUSTAINABLE VISION AND MISSION

This can guarantee that regardless of whether systems and practices change, the association stays submitted toward conveying a positive change to the earth, the general public, and inner and outer partners. A sustainable association makes sustainability a "part of its DNA" or corporate personality.

HR can assume a basic part in encouraging advancement of the association's sustainable vision and mission that affects groups, employees and different partners.

Part of HR in making a reasonable vision, mission, and technique:

1. Facilitate improvement of a sustainable vision, mission, and technique.
2. Assess if the current vision and mission tend to every one of the segments of a sustainable approach.
   - Develop a convincing business case for fusing sustainability in the association's vision and mission to top initiative by acquiring an outside vision – in context. This should be possible through benchmarking and contemplating sustainability objectives in different businesses and featuring how these have helped businesses in pulling in, drawing in and holding ability, in this manner, prompting enhanced business.
   - Adopt a cross-functional approach amid the business arranging procedure to guarantee that the business technique is reasonable by inciting chiefs on key sustainability goals, for example:
     - Long-term affect on condition and society.
     - Long-term affect on inside and outside partners, for example, clients, workers, and store network.
3. Make bolster for sustainability technique by getting worker viewpoint.
   - Provide contribution on employees' points of view on sustainability, for instance, how they figure the association can add to the security of nature and advancement of society. These points of view can be valuable while drafting the sustainability methodology of the association.
4. Impart the vision and mission overall levels in the association:
   - Work evenly and vertically over the association. Assume a main part in spreading the sustainability vision and mission at all lev-

els of the association.

- The sustainability vision can't be achieved in the event that it isn't drilled by employees as a feature of their part.

19.1.1. Aligned Leadership

HR can guarantee that employee activities get connected to sustainability by making green occupations, consolidating and adjusting worker key result area (KRAs) or key performance indicators (KPIs) to the sustainability objective.

Adjust center procedures, for example, enrollment and determination, preparing and improvement, execution administration, prizes and acknowledgment to the association's sustainability logic.

19.2. BUILDING AN ALIGNED LEADERSHIP PIPELINE

Groundbreaking business pioneers comprehend the significance of sustainable business practices for the long haul eventual fate of the association and society. The authority drives change in an association by setting up heading, adjusting fitting assets and moving through case. Pioneers go about as champions by taking up and supporting the reason for sustainability in the association.

19.3. ROLE OF HR IN BUILDING AN ALIGNED LEADERSHIP PIPELINE

Adjust administration to the sustainability plan:

- An essential viewpoint of sustainability is assorted variety – guaranteeing that diverse perspectives are taken and assessed. HR can make suitable discussions together. This can unite various perspectives and diverse ranges of capabilities on a solitary stage and connect with the initiative on the sustainability plan.
- Communicate and hoist the need and significance of sustainability at each accessible gathering to feature the effect of reasonable business objectives on the group, the clients and on key business parameters.
- Align the execution administration system at an authoritative level and at an individual level through KPIs, including those

identified with economical development. This is to guarantee that feasible execution is recorded, followed, checked and considered for execution assessment in the meantime.

- Align the prizes system to drive feasible reasoning. Progressively, businesses are driving economical practices by adjusting short-term and long haul motivations as a major aspect of official remuneration. Normally, businesses use incentives, for example, value settled Long-Term Incentive Compensation (LTIs) and conceded reward to drive long haul thinking toward business comes about.

Building a reasonable administration pipeline by standardizing initiative advancement frameworks:

- Evaluate the current competency system to check its suitability and fuse the basic administration capabilities to drive and execute sustainable business objectives. Consideration ought to likewise be paid to cutting edge skills expected to drive the business system. A suitable competency demonstrating activity might be expected to adjust and change the current structure.
- Vital for fruitful execution at the part and in driving sustainable business practices.
- Organizations direct evaluation and improvement these skills and recognize holes and territories of advancement. Progressively, businesses are utilizing methods, for example, Multi-Stakeholder Feedback (MSF) and psychometric instruments alongside advancement and appraisal focuses to get a complete comprehension of improvement zones.
- Identifying the improvement territories is just a section of the general administration arrangement and advancement process. HR experts need to guarantee advancement of these skills through the creation and time-bound finishing of Individual Development Plans (IDPs). A few businesses utilize the 70–20–10 standard for production of IDPs. The standard suggests that 70% advancement happens at work, 20% from instructing and feedback and 10% from classroom preparing. Businesses are utilizing diverse advancement mediations including Management Development Programs (MDPs), part pivot, extend assignments, training, and coaching to address improvement needs.
- Identifying fitting successors for key positions of authority to

drive the business system and the sustainability motivation. This should be possible by assessing the status of every pioneer through evaluation of and abilities.

19.3.1. Creation of a Suitable Leadership Pipeline

Production of suitable association structure and administration systems to drive the sustainability plan.

As sustainability keeps on going up in the need list for the C-suite, businesses are finding a way to drive feasible business objectives crosswise over levels and groups. Be that as it may, businesses regularly neglect to execute sustainability and insufficient power and assets allotted for driving the sustainability plan.

The association needs to guarantee that the sustainability motivation stays on track by distinguishing objectives and targets, checking advancement and revealing results. Without a formal structure measure accomplishment in the usage of sustainable business objectives.

With a specific end goal to accomplish enhanced lucidity and development, businesses:

- How would we implant sustainability into the esteem chain and metrics?
- How would it be a good idea for us to adjust the sustainability technique to business procedure?
- Should there be a different capacity/group to drive sustainability activities?
- Where in the association's design should this group dwell? In the life structures of a feasible association, the structure frames the spine. Contingent on the development of sustainability activities in the association, the duty regarding them can be dispensed in various routes as point by point in Figure 19.2.

Once a reasonable structure is set up; it must be supplemented with fitting administration systems through proper appointment of expert to empower usage and basic leadership.

Part of HR responsibilities is to ensure a proper structure to drive the implementation of sustainable business practices.

Option 1: Centralized Model

This model treates sustainability as the sole responsibility of a separate department. The model creates a dedicated focus on sustainability and also helps in developing specialized skills in this area.

Option 2 : Decentralized structure

This model treats sustainability as the agenda of various business units. Sustainability officers report to business heads as well as a executive level position at the company.

Option 3 : Integrated structure

This model does not involve creation of separate functions. Rather, it treats sustainability as part of everyone's role . Additionally, specific roles in each department double up to drive sustainable business practices.

Figure 19.2: Structural options for sustainability in organizations.

Source: Ernst & Young (2017)

Encourage the creation of a suitable organization structure:

- The HR work assumes the part of a practical master in comprehension and making the most reasonable association structures, making suitable part depictions and distinguishing KRAs and KPIs by picking a fitting model for association plan. Figure 19.2 delineates EY's Organization Design Model.

Be that as it may, the part does not end at that. The most sustainable arrangements are the ones that develop continually. This requires the HR capacity to assume the part of a reception apparatus, which persistently pays special mind to new advancements in the earth. Similarly, as the lifecycle of the business moves and the association develops so will the level of its sustainability activities. The HR work should be mindful of the effect of these progressions and predict the requirement for auxiliary adjusted to the sustainability plan.

- Selection of individuals to possess the parts in the association structure is similarly basic, if not more, for effective usage of the sustainability plan.
- Creation of suitable administration instruments: HR can assume a basic part in recognizing key members in administration advisory groups. They can work crosswise over capacities to guarantee satisfactory portrayal of every single basic capacity and organizations (Figure 19.3).

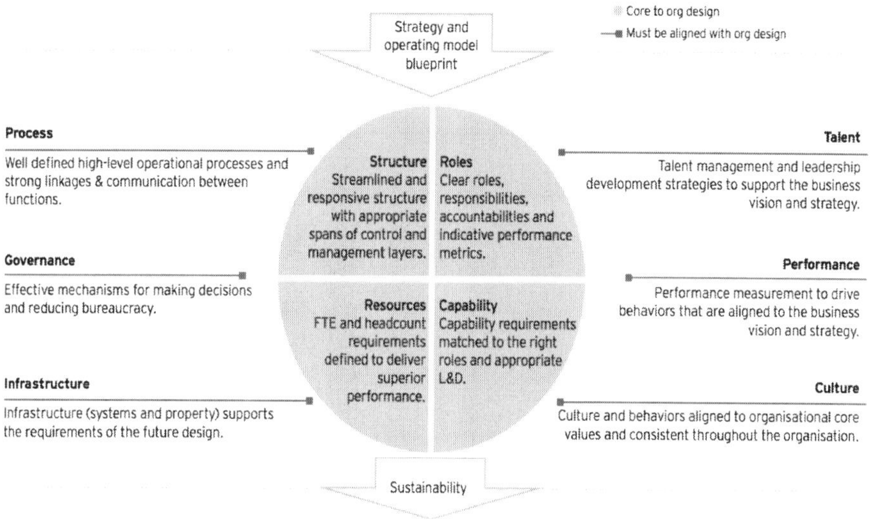

Figure 19.3: Organization design model.

Source: Ernst & Young (2017).

19.4. DEVELOPMENT OF KEY HR CAPABILITIES TO DRIVE SUSTAINABILITY AND CREATE SUSTAINABLE HR SYSTEMS AND PROCESSES

To impart the sustainability vision and mission, analyze the current capacities of the association to actualize reasonable business practices and drive change toward sustainability attitude, HR must be outfitted with the abilities to do as such. The capacities of the HR capacity can be landed at by assessing key parts and duties that should be performed and the hidden skills expected to convey these obligations as delineated in Figure 19.4.

Do

Development of:

Role objective

- Baseline responsibilities
- Reporting relationships
- Key interfaces (internal and external)
- Qualifications
- Experience levels

Display

Identification of:

- Technical and behavioral competencies required for the role

Derived from the role competency profiles

Deliver

Definition of:

- Key Result Areas (KRAs)
- Key Performance Indicators (KPIs)

Derived from the Performance Management System

Figure 19.4: A three-dimensional job description model.

Source: Ernst & Young (2017)

In our view, HR assumes the accompanying key parts to drive sustainable business objectives in businesses:

- Collaborator significance of sustainability crosswise over pioneers, levels, and groups by featuring best practices on sustainability and key.
- Orchestrator to make suitable gatherings to get pioneers crosswise over capacities together to talk about, ponder and figure a business case to make a dream, mission, and system, which is sustainable.
- Moderator to direct and incite senior authority amid the business arranging procedure to guarantee that business heads make an economical technique.
- Assessor to survey and analyze the condition of the association to drive change toward a sustainability mentality.
- Business accomplice to see how business impacts sustainability and help make a motivation, which is a win-win for both the business and the general public on the loose.
- Communicator to convey the sustainable vision and mission overall levels and capacities in the association.

- Change specialist to drive change in the practices of the workers at all levels toward a sustainability mentality.
- Functional master to have the capacity to make frameworks of tomorrow today by inserting sustainability as an outline rule into HR frameworks and procedures.

Key HR capabilities needed to deliver on the above roles and drive sustainability in the organization.

Business information: understanding sustainability with regards to the business:

- Understand sustainability, the key objectives and parts of a reasonable approach and how sustainable business objectives
- Develop a profound comprehension of business including the association's plan of action, its esteem chain, clients, providers, and operations to break down the different touch-focuses with the earth and society and distinguish opportunities for making a positive effect by working crosswise over levels and capacities.
- Joint effort and cooperation: working together crosswise:
- Collaborate with initiative crosswise over various capacities to make an economical vision, mission, and technique.
- Work crosswise over levels to convey and drive the sustainability motivation.
- Distinctive levers to take into account diverse partners Functional capabilities: assemble practical HR frameworks and procedures.
- Develop capacity to work forward-looking HR frameworks and procedures to help recognize key relationship building abilities and abilities for effectively executing the vision of the association.
- Develop HR frameworks and procedures, which are conscious of the desires of age Y and the future workforce to guarantee the association keeps on pulling in and hold ability with the evolving.
- Change administration: capacity to drive authoritative change:
  - Ability to survey the preparation of the association for change.
  - Ability to make a change vision and drive association change.

Capacity to draw in, connect with and hold the correct arrangement of individuals and adjust them to the sustainability motivation

A basic part of sustainability is the capacity to draw in and hold the correct ability to convey forward the vision into what's to come. Moreover, with the approach of online networking, ability is winding up progressively aware of the effect of businesses on the earth and society on the loose. Along these lines, right practices.

The business mark and related Employee Value Propositions (EVPs) address a wide scope of components of the "arrangement," which matters to employees. Businesses, which are hold ability on a proceeded with premise (Figure 19.5).

Figure 19.5. Key component of EVP.

Source: Ernst & Young (2017)

Keeping in mind the end goal to make an economical workforce pool, the HR work must assume a basic part in empowering the same by developing a hearty arrangement of EVPs which endeavors to adjust worker's desires and association's desires.

Businesses need to implant components of the sustainability motivation into EVPs to meet the desires of future and existing workers.

A sustainable incentive for employees should address the accompanying:

- Align with the sustainable vision and mission of the organization.
- Maintain adjust between worker desires on vocations, prizes, and improvement on one hand and capacity of the association to convey through reasonable duties and desires setting.
- Empower employees with significant work and capacity to strike

work-life adjust.

- Create long haul vocation opportunities for employees instead of just occupations.

The HR capacity can assume a basic part in making/looking into EVPs and conveying them to workers:

19.5. HR PART: OVERSEEING HIERARCHICAL DESIRES

Adjusting employees at all levels to vision and mission and procedure: Organizations expect that their workers' objectives are adjusted to the association's vital goals.

Moreover, they anticipate that workers will disguise the capacity assumes a key part in this perspective by encouraging the falling of general business objectives to employees as far as KRAs and KPIs. It can likewise help in conveying the association.

19.6. HR ROLE: MANAGING EMPLOYEE EXPECTATIONS

Making and imparting the business brand: Every singular wishes to work for a business with a great notoriety, and is regarded by inner and outside partners. Regularly, bosses improve the items and administrations, affect on condition and society and by making a positive association with interior and outer partners.

Nonetheless, it is similarly imperative to impart a solitary message featuring the brand and its key components both inside and remotely. HR can do this inside through suitable correspondence channels.

- **Creating Employees to Upgrade Their Abilities for the Future:** Employees hope to build their insight and aptitudes while working for an association. They trust that businesses ought to give formal preparing projects to them to address any expertise holes. The HR capacity can assume a key part in recognizing aptitude holes, assuming any, in the association and devise preparing programs that can grant vital abilities. Be that as it may, most businesses just concentrate on current aptitudes and not modern abilities required to convey the vision and mission. For example, a global inn network trains inn administration understudies by turning them crosswise over various branches under a two-year program to build up their authority aptitudes.

- **Making Work That Connects with and Inspires Employees:** Employees have a solid inspiration to work when they and look to have any kind of effect. They look for a part that is testing and can give them a solid feeling of self. The work part ought to likewise permit innovativeness, assortment, learning, and self-sufficiency. In such manner, HR can assume a key part by making nitty-gritty sets of responsibilities, coordinating the perfect individuals with the correct part and setting worker desires right particularly at the season of on-boarding. Employees can likewise be urged to take part in CSR exercises as a component of their part necessities to add all the more importance to their work.
- **Overseeing Singular Profession Desires:** Often, vocation objectives are the end toward which a person's exertion is coordinated. These objectives express what people want to accomplish inside their vocations and steer profession related decisions, which impacts their maintenance in the association. Sustainable businesses need to comprehend singular vocation objectives. The HR work helps in bringing clearness around singular vocation objectives and in creating objectives. Besides, from a framework point of view, HR must guarantee that worker tried parts and pace of development are aligned with an association's ability to give the same.
- **Making Fair Rewards and Acknowledgment:** Compensation is an imperative factor to pull in and retain workers. Workers are most worried about being reasonably remunerated for the additional exertion they put in.

This helps support employee resolve and furthermore urges different employees to go the additional mile. HR can help match remuneration to execution. It can likewise help in understanding the manner by which an employee's value is remunerated. A few employees may esteem pay over factor pay. HR can guarantee that the association's prizes and acknowledgment program urges workers to show proper practices. Reward frameworks need to keep up inner and outside value.

In accordance with the long haul center of a practical association, the prizes framework should likewise be intended to make long haul an incentive for the worker. There is a need to mix group and individual impetuses for enhanced sustainability, as more joint effort is required. Acknowledgment plans need to move past the association and include accomplices and partners, given an expanding center around biological communities.

19.7. CREATING A WORK ENVIRONMENT TO IMPROVE PERFORMANCE

A sound and safe workplace enable employees to build up a compelling responsibility toward the business. A superior workplace gives a feeling of prosperity made by the physical surroundings of the work environment. The organization should deliberately cultivate a culture that incorporates with its key goals. For instance, employees at a main American multinational organization gaining practical experience in web related administrations and items have a day-by-day employment of thinking of creative arrangements, and what gets this going is the interesting working style that empowers advancement and solid negligence for the unimaginable. Other than concentrating on wellbeing and security in the work environment, bosses need to guarantee that the workplace regards assorted variety.

- Facilitating work-life adjust: Employees wish to strike a harmony between their own and working existence without trading off on either. Organizations, which can offer exceptionally energetic and high-performing employees, a better harmony between their own and expert life can pull in and hold the best ability. Offering work – businesses can help employees to accomplish work-life offices, for example, day care and day-by-day require administrations, for example, cleaners and auto repairs.
- Apart from making a convincing EVP and keeping up the harmony amongst association and employee desires, HR can help in connecting with workers to take an interest in natural and social exercises.

Formation of a lithe association ready to learn and adjust to changes activated by the interior and outside condition.

Lithe businesses are sustainable as they can adapt to changes activated by the inner and outer condition. Production of a deft association requires the accompanying two abilities:

- Ability of businesses to gain from the earth and past encounters.
- Ability to rapidly adjust to changing interior and outer condition.

- A capacity of businesses to gain from past encounters and condition.

"An association's capacity to learn, and make an interpretation of that learning vigorously quickly, is a definitive upper hand"

—Jack Welch

At the point when characterized regarding environment, a sustainable organic framework is one, which stays assorted and profitable after some time. For businesses, sustainability is the long haul support social, natural and monetary requests. While numerous businesses figure out how to accomplish this compromise for quite a while, a proceeded with accomplishment of this objective requires continuous change and change. With a specific end goal to be versatile to change and grow, an association must guarantee that it is completely prepared to do as such regarding learning and abilities.

This creates a need to develop a "Learning Organization."

"A learning organization is an organization skilled at creating, acquiring and transferring knowledge and at modifying its accompanying changes in the way that work gets done; only the potential for improvement exists"

—David A. Garvin

While numerous businesses take activities to make a learning association, just a couple of genuinely accomplish a level where learning is organized in frameworks and procedures. While the ideation around learning is set up, its ground level execution is the place the issues hold on.

A portion of the difficulties, which are looked at the ground level include:

- Lack of information among employees on accessible frameworks and procedures for learning sharing.
- Time and cash limitations.
- Lack of supporting frameworks, (e.g., learning administration entrances).
- Lack of instruments for evaluating whether groups are learning and how this picking up helping the association is.

19.8. PART OF HR IN MAKING A LEARNING ASSOCIATION

- Formalizing the procedure of casual learning and information sharing.
- Facilitate formation of proper stages and discussions for sharing learning and best practices crosswise over association by working with administration crosswise overcapacities.
- Allocate suitable time for learning administration activities over the association and fabricate it as a feature of key duties of every part in the association.
- Management KPIs are embraced by every employee to drive information sharing.
- Train line chiefs on advancing a learning society in their group's capacity to rapidly adjust and change to changing interior and outer condition (Figure 19.6).

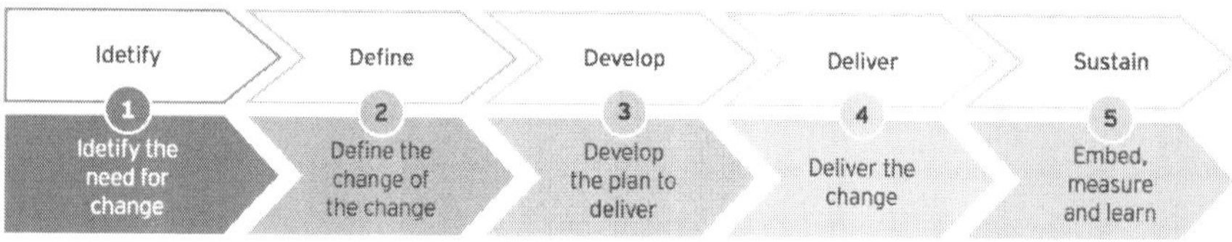

Figure 19.6: Change management framework.

Source: Ernst & Young (2017)

Sustainability is a coveted future state for businesses. Be that as it may, the push toward sustainability includes a change in authoritative and employee outlook. Furthermore, a practical association should be nimble to learn and adjust quickly to changes in the outer condition. HR professionals can drive the change administration process in a sustainable association given their one of a kind and process-situated ranges of abilities for driving authoritative change.

19.9. PART OF HR IN CHANGING WORKER PRACTICES OVER THE CHANGE MANAGEMENT LIFE CYCLE

The HR capacity can help drive the adjustment in worker conduct from attention to responsibility regarding a sustainable method for working.

- **Identify:** To change worker practices, the HR function can help in making a case for sustainability by featuring its effect on the association. For this situation, it should be featured that the effect may not always be monetary. Or maybe it can be regarding helping its ecological and social obligations. They have to secure sponsorship from administration to drive the change in practices. This implies it ought to build up a convincing vision with clear and achievable sustainability targets.
- **Diagnose:** HR experts can likewise help in understanding the change toward sustainability can likewise be surveyed in this stage.
- **Design:** HR can evaluate the change effect and configuration connects holes and oversees change affect. Mediations need to concentrate on drawing in and dealing with all partners, adjusting individuals, structures, procedures and frameworks to a more sustainable method for working.
- **Deliver:** HR experts can assume a basic part in the real conveyance of outlined change administration designs. HR experts can go about as good examples to encourage the change toward sustainability, propel and draw in with workers to deal with individuals' trip through the period of progress. At the general population level, HR should be associated with the enrollment, learning and advancement and maintenance of employees. At the partner level, HR can convey and accumulate input to guarantee operational business availability to change.
- **Sustain:** All together for a change to support in the association, HR can make checking mechanisms to survey if the change has been successful. It needs to celebrate with employees the better approach for working and the new sustainability mentality. It is additionally vital at this crossroads to record refreshed frameworks, process and arrangements and make learning administration systems to guarantee data is open to existing and new employees on the better approach for working.

19.10. GLOBAL WORKPLACE CHALLENGES

19.10.1. State of the Global Workplace

Right now, 13% of workers crosswise over 142 nations worldwide are engaged with their employments – that is, they candidly put resources into and concentrated on making an incentive for their businesses consistently. As in Gallup's 2009–2010 worldwide investigation of employee engagement, effectively withdrew laborers – i.e., the individuals who are negative and conceivably unfriendly to their businesses—keep on outnumbering engaged workers at a rate of almost 2–1.

Worker engagement will turn into an inexorably imperative worry for nations and businesses trying to help work efficiency as the worldwide economy proceeds with its quick pace of progress. Indeed, even as joblessness rates stay high in many developed world nations five years after the beginning of the worldwide monetary emergency, development rates in China and other huge developed nations have been falling.

Around the globe, statistic patterns are having significant monetary results. In a few regions – including southern Europe, South Asia, and the Middle East – an "adolescent lump" keeps on delivering record joblessness among youngsters. In the interim, numerous expansive economies – including China, Japan, and the U.S. – confront ability deficiencies as their workforce's age.

One normal factor among businesses worldwide is the need to all the more adequately comprehend and utilize their core abilities, aptitudes, and energy. In numerous nations, raising specialists' profitability levels is basic to business development and gravely required occupation creation. In nations that face ability deficiencies, organizations that address their workers' issues are destined to win in the opposition for top ability. For each situation, organizations around the globe should enhance their capacity to guarantee that specialists are in the correct parts and are sincerely put resources into their employments.

At the end of the day, the need to construct very engaged work environments will turn out to be more essential than any other time in recent memory. Through many years of research with several businesses and more than 25 million employees, Gallup has built up an unparalleled comprehension of what the world's most grounded businesses do any other way and how engagement influences profitability and worker prosperity in any work environment.

19.10.2. Engagement Makes a Difference to the Bottom Line

- Engaged laborers are the soul of their businesses. Work units in the main 25% of Gallup's Q12 Client Database have fundamentally higher efficiency, gainfulness, and client evaluations, less turnover and truancy, and less wellbeing episodes than those in the last25%.
- Organizations in Gallup's Q12 Client Database with a normal of 9.3 engaged employees for each effectively separated worker in 2010–2011 experienced 147% higher income for every offer (EPS) contrasted and their opposition in 2011–2012. Conversely, those with a normal of 2.6 engaged workers for each effectively withdrew employee experienced 2% bring down EPS contrasted and their opposition amid that same day and age.
- Active withdrawal is a massive deplete on economies all through the world. Gallup gauges, for instance, that for the U.S., dynamic withdrawal costs US$450 billion to $550 billion every year. In Germany, that figure ranges from €112 billion to €138 billion every year (US$151 billion to $186 billion). In the United Kingdom, effectively separated employees cost the nation between £52 billion and £70 billion (US$83 billion and $112 billion) every year.

19.10.3. Engagement Levels Vary Widely From Region to Region

- Overall, among the 142 nations incorporated into the present Gallup consider, 13% of employees are engaged with their employments, while 63% are not engaged and 24% are effectively maintained. Be that as it may, these outcomes change considerably among various worldwide areas.
- East Asia has the most reduced extent of engaged employees on the planet, at 6%, which is not as much as half of the worldwide mean of 13%. The territorial finding is driven transcendently by comes about because of China, where 6% of employees are engaged with their employments – one of the most reduced figures around the world. China's low engagement level may progressively represent a hindrance to its proceeded with development as the nation makes the change to a more buyer based economy and organizations come to depend more on cutting edge workers to

pull in and hold clients.

- In Australia and New Zealand, 24% of workers are engaged, while 60% are not engaged and 16% are effectively withdrawn. The subsequent proportion of connected with to effectively withdrew workers – 1.5-to-1 – is one of the most noteworthy among every worldwide area and like outcomes from the U.S. also, Canada (1.6-to-1).
- Gallup found the largest amounts of dynamic separation on the planet in the Middle East and North Africa (MENA) area, especially in Tunisia (54%), Algeria (53%), and Syria (45%). The district's high joblessness rates might be a factor in these outcomes, causing many withdrew specialists to stay in their employments in spite of their despondency at work.
- Despite the nation's solid financial development, just 8% of Indonesian employees are engaged with their employments, while 15% are effectively withdrawn. By differentiate, workers in the Philippines – another quickly developing economy in Southeast Asia – had the most abnormal amount of engagement in the locale at 29%, with just 8% effectively separated.

19.10.4. In the Wake of the Global Recession, Engaged Employees Are More Likely to Report Job Growth in Their Organizations

- Organizations worldwide that augment employees' inspiration and eagerness for their work are thus animating employment creation. All inclusive, 44% of engaged employees say their bosses are enlisting individuals and growing the span of their workforces vs. 34% of the individuals who are not engaged and 25% of effectively separated workers who say the same. This relationship is reliable over every single worldwide locale.
- Perceptions of occupation creation are most reduced internationally among employees living in the Commonwealth of Independent.
- States and neighboring nations. In general, 24% of workers living in this gathering report that their managers are employing new individuals, the most reduced figure among every local gathering. Almost four out of 10 connected with specialists say their bosses are growing the span of their workforces, contrasted and around

two out of 10 not engaged or effectively withdrew employees.

- One-third of workers in the MENA locale (33%) say their bosses are procuring new individuals – an assumption that should ascend in the coming time to meet the district's expanding requirement for new occupations to employ its blossoming youth population. Over the MENA locale, about portion of connected with employees say their businesses are employing versus somewhat more than one-fourth of effectively withdrew employees who say the same.
- Job creation is no place more vital than in South Asia, where it is assessed that more than a million people will enter the work drive each month in coming years. At present, employees working in South Asia are about as prone to state their businesses are releasing individuals (25%) as they are to state their managers are employing (28%). Among engaged employees, be that as it may, more than four of every 10 (42%) say their organizations are employing, while 12% say they are releasing individuals.
- Many nations in Western Europe, including France, Ireland, Italy, and Spain, keep on suffering from serious business emergencies. Occupants in these nations are among the most outlandish on the planet to state it is a decent time to discover a vocation in their groups. Be that as it may, connected with employees are twice as prone to report their working environments are contracting (34%) as they are to state they are releasing individuals (16%). Among effectively separated workers, these figures are turned around, with 34% saying their businesses are releasing individuals, while 18% say they are enlisting.

19.10.5. Poor Hiring and Management Practices Hinder Companies' Growth and Engagement Levels

- In the MENA locale, the idea of "favors" (like the Western idea of "who you know") can be utilized to pick up business and favorable position in the work environment, because of individual connections, and possibly undermine working environment engagement. Not exclusively does this framework make poor fit for parts among shamefully contracted employees; however, it likewise spreads negative discernments among generally connected with partners.

- In quickly developing Southeast Asian economies like Indonesia and the Philippines, where interest for work is high, organizations will require a solid ability strategy to frustrate ability poaching. Vital to the accomplishment of such systems will be administrators' readiness to shun the old "order and control" mindset for a more shared approach, especially with more youthful laborers.
- In East Asian social orders, the social estimation of deference to expert may make organizations more averse to focus on administration structures that permit workers to feel equipped for stepping up. Just around one out of six workers in East Asia emphatically concur that their conclusions tally, the most minimal extent of any worldwide locale.
- The positive energy in the Latin American job showcase has uplifted the opposition for talented specialists among organizations in the region. The force has likewise expanded the significance of building devotion among workers by guaranteeing that they have what they should be completely engaged with their occupations.
- In Australia and New Zealand, just 19% of workers in authority positions are engaged with their jobs. Low engagement among administrators is alarming for organizations, as Gallup has discovered that they assume the most noteworthy part in impacting engagement among their immediate reports.

19.10.6. Worldwide, Engaged Employees Regard Their Lives More Highly and Experience More Positive Emotions

- Around the globe, connected with workers will probably be "flourishing" – i.e., to rate their general lives highly on a zero-to-10 scale – than the individuals who are not engaged or effectively separated. Among every worldwide locale, engaged specialists are no less than 1.6 times as likely as effectively separated laborers to flourish. The proportion is most elevated in South Asia, where connected with workers are 5.5 times as liable to flourish as the individuals who are effectively separated.
- In East Asia, engaged specialists are about half as liable to have encountered pressure the earlier day as their effectively withdrew peers. Earlier research offers some proof that those in East Asian social orders are more averse to draw on social help in distressing circumstances. Therefore, working environment factors – and

great chiefs specifically – may have to a greater degree an effect in guaranteeing that employees don't hesitate to request oversee pressure all the more viably.

- Latin America has the most noteworthy rate (55%) of flourishing employees in any creating locale. The association with work environment engagement is likewise especially solid there: 66% of engaged employees (66%) are flourishing, contrasted and not as much as half (42%) of effectively separated laborers.
- About 33% of MENA occupants (34%) say they encountered outrage for a great part of the day preceding the study, effortlessly the most astounding extent among any worldwide locale. Be that as it may, this number tumbles to 19% for engaged employees in the district, contrasted and 35% of their effectively withdrew partners. Similarly, 75% of engaged employees versus 48% of effectively separated workers experienced satisfaction "yesterday."
- Similarly, in sub-Saharan Africa (SSA), connected with employees are essentially more probable than effectively separated workers to state they encountered delight for a significant part of the day "yesterday" and about half as liable to have encountered outrage.
- The seriousness of Western Europe's obligation emergency and the going with gravity measures in numerous nations may help clarify why life assessments in the greater part of the area have a tendency to be lower than those in other developed districts. Half of employees in Western Europe (half) are flourishing in their general lives, contrasted and 59% in the U.S. what's more, Canada and 66% in Australia and New Zealand. Among connected with employees in Western Europe, the flourishing rate rises to some degree to 59%.

19.10.7. Education Is Often Associated With Higher Engagement Levels – If Employees Can Find Jobs That Make Use of Their Knowledge and Talents

- In many creating and transitional economies where the interest for high-gifted work surpasses the supply of all the more exceedingly taught occupants, better-instructed employees will probably be engaged at work. In any case, this isn't really valid in developed economies, a large number of which are enduring an expanded time of high joblessness. In the U.S., furthermore, Canada, for instance, engagement slants descending somewhat with workers' training level. School taught laborers in the two nations are more outlandish than those with just a secondary school instruction to firmly concur that they have the chance to do what they excel at work everyday.

- Likely mirroring the confound between instructive educational program and the necessities of managers in the district, one of every four MENA employees with a school level training (25%) are effectively separated. This is the most astounding extent among exceedingly instructed laborers in any worldwide district.
- Among employees in Southeast Asia, there is an especially solid connection between workers' instructive achievement and their engagement levels: Those with a basic training or less are right around four times as likely as those with a school training to be effectively withdrawn at work. Those in occupations more averse to require larger amounts of formal training are likewise most drastically averse to be engaged at work, exhibiting a requirement for organizations in mechanical divisions to enhance correspondence and acknowledgment frameworks that assistance rouse and enable individual laborers.
- Similarly, engagement rates increment with Latin American employees' instructive fulfillment levels. The engaged effectively separated proportion is most good in work composes that have a tendency to require more elevated amounts of training and give a more independent workplace, for example, proficient laborers and those in administration or positions of authority.

19.11. WHAT COMPANIES CAN DO TO IMPROVE ENGAGEMENT?

- **Carry Engagement Into the Organization's Ordinary Dialect:** The organizations best at drawing in their employees bring the discussion of engagement into the working environment consistently. It is imperative before businesses start estimating engagement to impart the explanations for this vital objective and the favorable circumstances for the organization and the workers themselves. Normal correspondence from the organization's pioneers and casual correspondence between employees will start to breed a culture of engagement, driving support rates of worker engagement measurements and different intercessions to be more effective.

- **Utilize the Correct Worker Engagement Overview:** The worker engagement measurements organizations utilize can influence their capacity to make changes in execution. Frequently, businesses wrongly use employee overviews to gather information that are insignificant or difficult to follow up on. At the point when an organization approaches its employees for their suppositions, they anticipate that activity will take after. Gallup's Q12 employee engagement metric was composed in view of this desire the information the Q12 overview gathers are particular, significant, and noteworthy for any group at any authoritative level, and they are demonstrated to influence key execution measurements. Why? Since the Q12 measures workers' enthusiastic engagement, which binds straightforwardly to their level of optional exertion – their ability to go the additional mile for their organization.
- **Concentrate on Engagement at the Endeavor and Neighborhood Levels:** Change happens at the neighborhood level; however, it just happens when the tone is set starting from the top. Organizations understand the most advantage from engagement activities when pioneers mesh worker engagement into execution desires for chiefs and empower them to execute on those desires. Supervisors and employees must feel engaged by initiative to have a huge effect in their prompt condition.
- **Select the Correct Chiefs:** In the case of enlisting all things considered or advancing from inside, businesses that logically select chiefs for the one of a kind abilities it takes to successfully oversee individuals significantly increment the chances of drawing in their workers. Rather than utilizing administration occupations as limited time prizes for all profession ways, organizations should regard these parts as one of a kind, with particular practical requests that require a particular ability set. They should choose directors with the correct abilities for supporting, situating, enabling, and connecting with their staff.
- **Mentor Directors and Consider Them Responsible for Their Employees' Engagement:** Gallup's examination has discovered

that directors are basically in charge of their workers' engagement levels. Businesses should mentor administrators to play a dynamic part in building engagement designs with their workers, consider chiefs responsible, keep tabs on their development, and guarantee they persistently concentrate on sincerely captivating their employees. The top entertainers in Gallup's Q12 Client Database reliably influence employee engagement to some portion of their formal survey process, and most utilize these enhancements as a basis for advancements.

- **Characterize Engagement Objectives in Realistic, Ordinary Terms:** While the general association may set grandiose objectives for engagement, pioneers must make these targets important to workers' everyday experiences to breathe life into engagement. Guarantee that chiefs talk about worker engagement components at week after week gatherings and in one-on-one sessions with employees to mesh engagement into day-by-day cooperations and exercises identified with their execution targets.
- **Discover Approaches to Meet Workers Where They Are:** As this report illustrates, workers worldwide have distinctive necessities and desires that impact their engagement levels. Nearby social impacts and monetary conditions, and additionally more particular factors, for example, respondents' activity write and training level, all assume parts in molding their work environment encounter. Supervisors ought to know about the elements most pertinent to engagement among their specialists. They ought to likewise comprehend that each connection with an employee can possibly impact his or her engagement and motivate optional exertion.
- **What Good Work Looks Like:** Gallup examine finds that steady employments are what the world needs most. Worldwide overviews reliably exhibit that individuals with "steady employments" – defined as those that offer relentless work averaging at least 30 hours for each week and a paycheck from a business – are more probable than those in other business classes to rate their present and future lives decidedly. Subsequently, work creation is basic to a group's ability to address its inhabitants' issues, and pioneers worldwide must figure out how to draw in or hold skilled individuals who can make new organizations or enable existing ones to grow.

- Groups endeavoring to discover wellsprings of practical development require a "minimum amount" of high-energy organizations and businesses that reinforce work creation and financial open door. That energy will probably be found in working environments where employees are engaged with their employments. Drawn in employees are right around two times as likely as the individuals who are effectively withdrawn to report that their organizations are enlisting new individuals or growing. By differentiate, effectively separated employees report that their organizations are decreasing workforces at rates right around three times that of the connected with population.
- **Drawn in Employees Are More Optimistic about the Economy:** Gallup information proposes that connected with employees help cultivate business enterprise in their groups. Employment creation relies upon the certainty of a city's occupants to fire up new pursuits. Drawn in workers are more probable than their less connected with associates to believe in their groups' business condition: 71% of engaged employees report that their city is a decent place to begin a business, contrasted and 53% of the individuals who are effectively withdrawn. Drawn in employees are likewise more prone to concur that diligent work advances you beyond.
- **Connected Employees Rate Their Lives Better:** At the point when employees feel connected with and gainful at work, it emphatically influences their lives at work and past the work environment too. Connected with employees survey their general lives more exceedingly than not engaged or effectively withdrew workers. They likewise report more positive everyday passionate states and connections with others than their less engaged peers.

Every year in more than 140 nations and domains, Gallup's World Poll tracks two kinds of measures critical for surveying the subjective nature of individuals' lives: 1) their general life assessments, which are utilized to sort them as "flourishing," "battling," or "enduring," and 2) their particular passionate states, estimated by asking regardless of whether they encountered a progression of positive and negative feelings—including delight, stress, and outrage – the day before the study.

At the point when employees feel engaged and gainful at work, it decidedly influences their lives at work and past the working environment too. Connected with employees evaluate their general lives more exceedingly than not engaged or effectively separated workers. They likewise report more positive everyday enthusiastic states and connections with others than their less engaged peers.

Every year in more than 140 nations and domains, Gallup's World Poll tracks two sorts of measures vital for surveying the subjective nature of individuals' lives: 1) their general life assessments, which are utilized to classify them as "flourishing," "battling," or "enduring," and 2) their particular passionate states, estimated by asking regardless of whether they encountered a progression of positive and negative feelings—including pleasure, stress, and outrage – the day before the review.

1. **Life Assessments Depend on Engagement Perspective:** Striving Scale, with steps numbered from zero to 10. Flourishing respondents rate their present lives at 7 or higher, and foresee their lives in five years will be at 8 or higher. Enduring respondents rate both their present and future lives at 4 or lower.

The rest of those with midlevel appraisals – are named battling.

2. **Connected Employees are More Than Three Times as Likely to be "Flourishing" as Actively Disengaged Employees:** Worldwide in 2012, 31% of workers evaluated their lives sufficiently very to be viewed as flourishing, while the lion's share, 59%, were battling and 10% were enduring. Yet, these outcomes differed significantly as indicated by regardless of whether those workers were engaged with their employments. Connected with employees were more than three times as liable to flourish in their general lives as the individuals who were effectively separated.

This relationship is significant on the grounds that workers who are both engaged with their employments and flourishing in their general lives are more averse to be thrown off kilter by authoritative changes or disturbances in their own lives.

3. **Engaged Employees Have More Positive Daily Interactions:** Connected with employees likewise display more positive everyday feelings and encounters than not engaged or effectively withdrew workers. All engaged workers (95%) report being approached with deference the earlier day.

Connected with employees additionally encounter higher rates of satisfaction and lower rates of outrage and worry than their not engaged or effectively withdrew partners. Maybe most unsurprisingly, connected with employees are four times as likely as the individuals who are effectively separated to state they like what they do every day.

As the weight increments for group pioneers to provide steady employments and increment financial open doors, drawing in workers has never been more pertinent. Connected with workforces make outside advantages to the whole group through expanded monetary hopefulness among residents and enhanced execution results among organizations.

As Gallup CEO Jim Clifton notes in his 2011 book, The Coming Jobs War, "Fruitful nearby organizations are motors for work creation as well as motors for neighborhood social and group change." Through their impact on business results, endeavors to enhance employee engagement can possibly make the monetary energy important to manufacture more grounded groups

4. **Developing Markets Need Engaged Employees to Grow:** Quick development and advancement in developing business sector nations has drastically changed the world's monetary scene over the previous decade. The expression "developing business sector" has for some time been connected to the alleged BRIC countries of Brazil, Russia, India, and China. Be that as it may, financial specialists have all the more as of late-featured developing economic situations in various different nations, including Mexico, Indonesia, Turkey, and South Africa.

Much of the time, quick paced changes have accompanied developing agonies for these developing business sector nations. For example, some social parts advantage more than others do from financial development in these nations, bringing about extending wage imbalance. Gallup's overviews from a few developing business sector nations feature another factor that could be similarly as debilitating to the pace of development: low levels of employee engagement. Accumulated outcomes for employees in 23 developing markets uncover that around one of every 10 employees are engaged at work – about a large portion of the extent found crosswise over 23 developed advertise nations. (Nations are named developing markets and developed markets as per their present arrangement in the MSCI Developed-Market and Emerging-Market lists.)

An absence of worker engagement among organizations in developing business sector nations is probably going to wind up plainly an undeniably noteworthy issue. To maintain their development levels, these nations must depend less on ware generation and fares, and concentrate more on building up their household buyer markets.

Financial experts have noticed that the abating pace of development in many developing markets is a demonstration of this need to rebalance their economies. Developing buyer markets implies that organizations must know how to pull in and hold clients to keep up an aggressive edge in the commercial center. Unavoidably, these businesses will be those that comprehend and work as indicated by an essential standard bolstered by many years of Gallup examine: Engaged employees are the main ones who make connections with, faithful clients.

5. **Approaches to Accelerate Employees Engagement:** Following quite a while of research into human conduct in the working environment, Gallup found one of kind bits of knowledge and procedures to help organizations overall change the way they work together. To quicken engagement and advance growth within businesses, Gallup set up critical methodologies that pioneers in any nation or area can receive to boost the part human conduct plays in the working environment.

Finding the correct employees and supervisors is basic to an association's budgetary results. Gallup helps organizations logically select individuals who can possibly be top entertainers and backings businesses in contracting and advancing workers with the ability to support employee engagement.

It isn't sufficient to put the perfect individuals in the correct employments. Organizations must put resources into their employees' most prominent abilities to enhance their execution. Gallup enables businesses to find what is normally ideal with their employees by recognizing workers' one of a kind quality and after that expanding on them to enable employees to accomplish their maximum capacity.

At the point when utilized with a worker engagement activity, Gallup's exploration has demonstrated that putting resources into employee determination, qualities, and prosperity has the ability to help the outcomes organizations would get from expanding engagement alone.

6. **Employee Engagement Varies by Country:** While Gallup has discovered that 13% of employees are engaged around the world, there is extensive variety in engagement levels crosswise over various locales of the world, nations, organizations, and even among workgroups. This area gives a more intensive take a gander at the outcomes from Gallup's overall employee engagement contemplate by district, and much of the time by nation. The information uncover that the results related with employee engagement are reliable around the world. Crosswise over dis-

tricts with extensively differing social foundations and financial levels, employees who are mentally put resources into their work environments will probably report work development at their businesses, rate their lives higher all things considered, and encounter more positive feelings once a day.

Gallup's broad examinations and investigation demonstrate that factors, for example, an employee's instructive accomplishment level and kind of industry and occupation can represent numerous distinctions in engagement levels. The accompanying investigations exhibit how these distinctions play out in locales with various social attributes and at various financial levels.

In every nation, Gallup incorporated into the examination residents only who showed that they work for a business. In creating nations where many individuals work at casual individual or family-based undertakings, including subsistence-based exercises, for example, cultivating, less occupants addressed the employee engagement study things. Thusly, if less than 300 individuals addressed the employee engagement review things; Gallup did not report comes about at the nation level. For instance, this report gives nation level outcomes to four of the 32 sub- Saharan African nations studied, while in Western Europe, the report records comes about for every one of the 19 nations Gallup overviewed.

Whatever difficulties organizations look in a given country or locale – be they social, financial, institutional, or something else – their pioneers ought to perceive the significance of employee engagement as a factor that can enable them to keep up the confidence and diligence important for long haul achievement.

Pioneers who comprehend the significance of engaging and propelling their specialists in a work environment condition will be ready to beat their rivals and help give the basic monetary development the world needs.

Pioneers ought to perceive the significance of employee engagement as a factor that can enable them to keep up the idealism and tirelessness fundamental for long haul achievement.

CHAPTER 20

Talent Management

20.1. INTRODUCTION

Filled by globalization, innovative change, population development, and moving socioeconomics, sustainability superpowers will influence the capacity of business to succeed and flourish throughout the following 30 years. With the worldwide population anticipated that would increment from seven to nine billion individuals by 2050, organizations should rehash themselves to secure their entrance to assets and the social permit to work and develop. On the whole, mankind is as of now devouring 1.5 times the assets that can be restored in a year. Similarly as concerning, the pay imbalance hole keeps on broadening, additionally undermining social attachment. Organizations that foresee, anticipate, and help lighten these patterns won't just form their upper hand, they will add to building a suitable society for all.

For the business to succeed in this new worldwide setting, present and future pioneers will require a large group of new abilities and capabilities. Driving organizations will retool their ability administration frameworks to mirror the new competency objectives to position their organizations for future achievement. They will characterize the mission-basic ranges of abilities and outlooks pioneers need to make and secure future incentive for the organization and society, and install these in their authoritative competency models.

20.2. UNDERSTANDING COMPETENCY MODELS

Competency models are a key instrument to recognize, create, and convey top ability to drive superior and acknowledge business comes about. They characterize the execution and behavioral desires – the aptitudes and information – which add to business achievement. Driving organizations utilize competency Profiles to illuminate their ability administration techniques, including:

- Recruitment and determination;
- Onboarding;
- Training and improvement;
- Career improvement arranging;
- Job depictions;
- Performance administration;
- Compensation;
- Succession arranging.

Organizational competency models are basic devices for the determination and enrollment of up and coming pioneers or high possibilities. Indicating the key aptitudes and learning pertinent to authoritative achievement enables pioneers to comprehend what is required keeping in mind the end goal to enhance execution levels. By imparting reliable, detectable and quantifiable execution gauges, competency models give a guide to professional success. They additionally uncover opportunities for advancement mediations to upgrade individual and group competency levels. In addition, as businesses progressively remunerate directors not only for their outcomes, but rather for how they accomplish these outcomes, sustainability skills shed light on the attractive properties that legitimacy motivator pay.

Well-oversaw organizations overhaul their competency profiles because of mergers, an adjustment in the plan of action or another vital heading. Similarly, the changing business condition requires a competency display refresh. Previously, businesses worked inside a genuinely unsurprising, stable framework. Conversely, the present businesses work inside huge and complex reliant systems, expecting them to work all the more remotely and receive a more forward-looking methodology. With a specific end goal to encourage compelling basic leadership in this new business reality, customary administration abilities, for example, examination, relationship-building, and change initiative should be taken to another level. Business pioneers will require more noteworthy capability in overseeing complex foundational change and hierarchical change. They will likewise require further developed psychological, social and passionate limits.

20.3. ADDRESSING THE COMPETENCY GAP

Many driving officials accept there is a sustainability aptitude and learning execution whole in their businesses, as uncovered by the Ash ridge Business School in a 2008 worldwide review directed with 194 business pioneers. More than 75% think creating sustainability aptitudes and learning in their association is vital, while less than 10% trust their association is doing this well. They likewise trust that business colleges and expert enrollment bodies ought to convey sustainability instruction and preparing, yet few are doing this effectively. Tending to this begins with recognizing the sustainability skills that will be expected of future pioneers.

While much has been composed about this competency hole, there is an absence of direction on the most proficient method to address it through authoritative competency models. The present examination arranges the key abilities distinguished in the sustainability and administration writing esteemed basic to drive high sustainability and in this way authoritative execution and composes it into a competency display system.

More than 30 articles, inquire about papers and books distributed in the course of recent years by HR and corporate sustainability consultancies, sustainability business systems, administration specialists, scholastic organizations and others were checked on. From this, the best-prescribed abilities which are viewed as fundamental for business pioneers to guide their businesses through the dynamic sustainability landscape of the coming decades were distinguished. (See the Appendix for the rundown of distributions.) The underlying discoveries were imparted to specialists in the different abilities for approval and refinement. (The exploration directed for this report expands upon a buddy consider led by the creator of the corporate social obligation (CSR) criteria sheets should use in CEO enlistment and progression arranging, accessible here.)

20.4. SUSTAINABILITY TALENT MANAGEMENT

The literature review reveals there are five sustainability competencies – three skills and two knowledge areas – which companies should add to their existing leadership competency models to position the firm for future success. These include:

- Systems thinking;
- External collaboration;

- Social innovation;
- Sustainability literacy;
- Active values.

20.5. BUSINESS CASE FOR SUSTAINABILITY COMPETENCIES

While the sustainability super powers are new and their full impacts are not liable to be felt for at least 10 years, groundbreaking organizations would be astute to consider incorporating these five abilities in their authority profiles. The business advantages of doing as such include:

- Better capacity to foresee and oversee dangers;
- Increased advancement and opportunity distinguishing proof;
- Early access to new markets with economical items and administrations;
- Enhanced critical thinking and more successful basic leadership;
- Greater capacity to react to changing monetary conditions.
- Improved employee fascination, maintenance, and engagement.

Organizations that reskill their authority positions to prevail later on economy will discover early arrangements and understand an upper hand – while enhancing the social setting for their worker and client families and groups.

20.6. FIVE SUSTAINABILITY LEADERSHIP COMPETENCIES

For what reason does it make a difference?

Fundamental super patterns will influence the suitability of business and society and create expansive scale issues requiring all-encompassing, not reductionist, arrangements. Prompt difficulties are frequently profoundly installed in frameworks and must be tended to at the root through changing foundational cooperations, standards, or asset streams. A frameworks thinking viewpoint helps pioneers all the more completely comprehend the issues they confront, which thus encourages them to recognize use focuses to cultivate change. Numerous points of view increment experiences and alternatives to enhance hierarchical flexibility and deftness. They are the best way to completely comprehend complex issues and may permit win-win answers for surface notwithstanding when manage dug in struggle and rivalry. Frameworks to upgrade the capacity of pioneers to watch and distinguish behavioral subtle elements in the exercises of clients, providers, contenders, and others to cultivate and enhance advancement. This reasoning enhances ability to anticipate conceivable bounce back impacts – where picks up are made in one region with fall back in others.

20.6.1. Systems Thinking

What is it? It is the capacity to see and comprehend interconnections and association between performers (partners, connections, and so forth.) and areas (social, financial, ecological), crosswise over time (past, present to future) and scales (neighborhood to worldwide) to reframe and address difficulties and opportunities for business and societal advance.

What capabilities does it expand upon? Critical thinking, analytical thinking, and strategic planning.

20.6.1.1. Foundational

Applies frameworks thinking to issues and choices:

- Sees the master plan, takes a long haul to see and comprehends the expansive setting while tending to an issue or issue.
- Takes a comprehensive way to deal with the investigation that spotlights in transit a framework's constituent parts interrelate and how frameworks function after some time and inside the setting of bigger frameworks – can move from reality-based administration to design based administration as vital.
- Demonstrates "entire framework mindfulness" which is the ability to switch between alternate points of view, scales and perspectives to see the master plan and interconnections and interdependencies inside the framework at all levels, including having the capacity to downsize to little subtle elements.
- Frames issues as far as long-term patterns and underlying drivers as opposed to quick concerns.
- Constructs frameworks maps that find a test, association or strategy in connection to different performers, trends, organizations and so on that have made the present circumstance.
- Perceives the bigger framework and web of connections – businesses, supply chains, enterprises, urban areas, districts, and

eco-frameworks – of which the association is a section, instead of concentrating on the detail of a specific piece.

- Understands the interconnectedness and part of every on-screen character in the public eye (government, business, common society) and how they interface with each other.
- Understands the businesses between issues at a nearby, provincial, biological system, national or worldwide level, and creates business arrangements in connection to the scale and nature of the test.
- Pays regard for interrelationships, examples, and progression so as to search for interconnections, underlying drivers, and various causalities and to characterize and take care of issues.
- Takes a sweeping, 'superior' perspective of an issue or challenge and keeps away from discontinuity of subjects and over-rearrangements of issues as basically right/wrong or genuine/false.
- Sees businesses between expansive, divergent assortments of information.

20.6.1.2. Advanced

Applies frameworks thinking in joint efforts with a specific end goal to progress hierarchical and fundamental change:

- Works in inside and outside coordinated efforts to examine complex frameworks crosswise over industry segments, esteem chains, item life-cycles and geographic scales (neighborhood to worldwide) to see immediate and aberrant effects, exchange offs, falling impacts, inactivity, criticism circles, time delays and unintended outcomes of individual, hierarchical and aggregate activities to outline sustainable items and strong and versatile businesses and frameworks.
- Recognizes how changes to parts of the framework influence the entire and utilization a comprehension of how components of a framework relate and forms inside an entire associate to outline dangers and openings, distinguish conceivable use focuses or activities, upgrade them for a coveted result, make arrangements and advance the wellbeing of frameworks.
- Creates helpful change in money related markets, customer

conduct, hierarchical change, open strategy, innovation, and other substantial frameworks.

- Demonstrates consciousness of complex framework practices, for example, rise, causation, inactivity, and tipping focuses and the way they join to deliver unforeseen, non-straight and sudden change.
- Has aptitude in standards of frameworks administration, for example, flexibility and overseeing for rise.
- Applies multi-disciplinary and tri-segment (government, business, common society) points of view to framework change.
- Has huge multi-disciplinary experience, incorporating capability in at least two zones, for example, engineering and promoting, human sciences and modern outline, engineering and brain research.
- Uses framework thinking devices, for example, causal circle graphs, stock, and stream outlines and reenactment models in shared advancement activities.

20.6.2. External Collaboration What Is It?

It is the capacity to impact and lead change past business limits and crosswise over orders and fields. Coordinated effort can be between organizations, organizations and governments, organizations, and NGOs (non-administrative businesses) and multi-on-screen character organizations among each of the three. It empowers businesses to mutually create arrangements and cooperate for the soundness of the entire framework.

- **For What Reason Does it Make a Difference?:** The present hierarchical and societal difficulties are too huge and complex to be tended to by people or businesses working alone. Systems of interdependency make common dependence in which coordinated effort is vital for advance. An association's critical sustainability impacts are regularly found up or downstream of their operations, requiring communitarian exertion with providers, clients, business accomplices, contenders, and others to acknowledge arrangements.

Working cooperatively can produce unwaveringness and construct social funding to empower solid connections and cultivate authoritative flexibility. So as to investigate new market openings and shape the financial setting in which they work, businesses need to connect with others. Working crosswise over divisions and here and there esteem chains is turning into a business need as sustainability results must be accomplished through aggregate exertion that handles fundamental blockages crosswise over whole markets and purviews. It empowers an organization to open inventiveness and advancement and access different information, abilities, and systems while sharing danger and pool speculation.

1. **What capabilities does it expand upon?:** Inward collaboration, relationship building, and conflict resolution.

20.6.2.1. Foundational

Stakeholder issue recognizable proof and discussion:

- Understands the business case for viable outside coordinated effort – including opportunity costs (of teaming up versus not teaming up).
- Understands the key partners that effect and are affected by the association (emphatically and adversely, specifically or in a roundabout way) and comprehends their essential advantages, concerns, and desires; can portray accomplishment from the points of view of differing partners.
- Engages partners through meeting and discourse to recognize issues, resolve exchange offs, adjust contending requests and enhance items and administrations.
- Gains alternate points of view and experiences from systems administration and banding together with different people and businesses; adopts a key strategy to systems administration; comprehends estimation of systems for learning, information spread, and social change.
- Cross-prepares thoughts and cultivates choices and cross breeds by working crosswise over business orders and segments (e.g., creators and item engineers from one organization cross-treat with sustainability, deals and advertising specialists from another).
- Understands the parts of various businesses, to figure out who has basic leadership specialist and to build up administration, responsibility and correspondence measures.
- Understands issues at the convergence of business and society.
- Develops banding together and joint effort methodologies.

- Understands how to assemble communitarian exercises that advance value and horizontal reasoning in the decision-production process where every on-screen character contributes similarly, with a comparable level of engagement (i.e., sidelong rather than various leveled).
- Possesses abilities in advanced, web-based social networking and in-person correspondence and coordinated effort apparatuses and methods.

20.6.2.2. Advanced

Multi-partner co-made arrangements:

- Builds multi-party, activity arranged, commonly gainful coordinated efforts with differing divisions, enterprises, providers, clients, ages, disciplines, capacities, societies, groups, and businesses, including eccentric partners and contenders.
- Harnesses systems to achieve societal and frameworks change objectives.
- Fosters put stock in, commonality and shared comprehension to construct an aggregate vision, produce a typical purpose and encourage aggregate imagination to enhance business and social results.
- Understands and bridles various plans of action and societies from tri-segment involvement in business, government and common society.
- Navigates crosswise over various institutional and social limits and all divisions of society (counting academia, of all shapes and sizes social businesses, nebulous systems, instructive establishments, and so on.), drawing in with on-screen characters all through frameworks, incorporating people with significant influence and those in more negligible or defenseless positions in the public eye, to address foundational challenges.
- Understands and is touchy to power and benefit (i.e., has pondered social disparity, their own history and social position/rank – by and by, professionally, hierarchically, sectorally) and distinguishes and portrays key cases of how people and social gatherings encounter imbalance; outlines and executes procedures or medications that enhance the essentialness of energy and benefit.

- Understands how to address focused limits and rivalry related lawful issues.
- Leads virtual, cross-hierarchical groups.

20.6.3. Social Innovation What Is It?

It is the capacity to produce and empower plan of action, association and framework level development to progress both business and social esteem. It moves the aggregate concentration from receptive critical thinking to co-making the future with partners.

- **For What Reason Does it Make a Difference?:** How businesses respond to sustainability superpowers will influence their capacity to succeed and flourish. All segments, esteem chains, plans of action and capacities should be re-designed to empower synchronous business and societal advance.

Aptitudes in development age and usage are expected to acknowledge authoritative and substantial scale change. The capacity to transform many-sided quality into circumstance will give businesses a future upper hand.

- **What skills does it expand upon?**: Change Leadership, Creativity, and Innovation.

20.6.3.1. Foundational

Incremental bits of knowledge and arrangements:

- Identifies and actualizes chances to enhance and update items, administrations, and procedures to make incremental business and social esteem.
- Questions existing conditions, suspicions, aggregate standards and got knowledge; considers unheard of options and regards authoritative imperatives and discernments as discussable and transformable.
- Encourages a scrutinizing outlook and welcomes workers to challenge presumptions in view of past encounters; backings, remunerates and incorporates assorted and whimsical perspectives.
- Explores limitations in inventive ways and reclassifies difficulties to benefit from businesses crosswise over them; reframes issues as circumstances.
- Pushing ahead:

- Connects apparently inconsequential inquiries, issues or thoughts from various fields to develop new bits of knowledge.
- Experiments and gains from disappointment.
- Learns from inventive accomplishments outside the business and consistently examines cases from different ventures in group gatherings; remains side by side of sustainability drifts that are changing different divisions and thinks about how to apply them.
- Provides partners with open access.
- Makes choices when enough is known and fights the temptation to know everything; alters activities in light of what is found out.
- Confidently utilizes on the web and virtual instruments and assets to source and execute thoughts from outsiders.

20.6.3.2. Progressed

The association and framework change:

- builds and acknowledges optimistic dreams that encourage long haul authoritative and framework change.
- leads the advancement of, and conveys, a convincing future that creates social and business esteem; drives association wide activities to imagine and create answers for foundational issues and difficulties; outlines and oversees leap forward systems and plans in the midst of unexpected difficulties and impacts.
- shifts the aggregate concentration from responding to issues to co-making and building positive dreams for what's to come.
- encourages experimentation and development all through the association and deliberately utilizes fast prototyping systems to cultivate learning and assemble hierarchical energy.
- embraces vulnerability as an impetus for new thoughts and plans of action; utilizes versatile administration to progressively control complex change activities, changing the outline or system as the setting shifts.
- embraces imaginative pulverization as an open door, seeing astuteness in consider development of insecurity keeping in mind the end goal to cultivate transformative change; goes out on a limb to upset inheritance plans of action.
- encourages social business enterprise in the association and

empowers huge scale advancement; invigorates and augments the group of trailblazers; moves aggregate experimentation.

- Creates interdisciplinary groups and community procedures to develop sympathy through discussion, perception or encounters; utilizes human esteems and individuals driven methodologies, (e.g., human-focused plan techniques) to enhance or make new items and administrations.
- Surfaces, supports and deals with the dynamic strains of the association and partner framework and defies troublesome certainties about the present reality; deals with the imaginative pressure amongst vision and reality to move new methodologies and grow new systems; creates change procedures that distinguish unwanted states and elements and imagines attractive results:
  - Harnesses pressures as a wellspring of development and imagination.
  - Exploits openings introduce in different parts of the framework to make change around an issue by working crosswise over scales or authoritative/area limits.

20.6.4. Sustainability Literacy What Is It?

It is the learning of how ecological, monetary and social sustainability impacts the association and how the association impacts sustainability. It includes an attention to how to enhance the association's sustainability execution and how the association and plan of action can be adjusted to produce positive business and societal advantages.

- **For What Reason Does it Make a Difference?:** Sustainability powers will influence business and societal accomplishment over the medium to long haul. Foresighted organizations will use the sustainability patterns to plan new plans of action, construct their social permit to work and develop and emphatically impact societal results. Connected sustainability learning can improve esteem security and creation – and cultivate upper hand for the firm.
- **What Capabilities Does it Expand Upon?:** Business Acumen and Financial Leadership.

20.6.4.1. Foundational

Sustainability procedure and practical information:

- Familiar with foundational ideas of sustainability; learning of reasonable advancement and central socio-environmental ideas and standards.
- Aware of rising ecological and social patterns (counting little patterns and frail signs) and the dangers and openings they make for the association and its clients, providers, part and general business working setting over the medium to long haul.
- Knowledge of the association's immediate and circuitous social and natural effects.
- Aware of how partners and their associates are reacting to sustainability uber powers and can distinguish and organize important partner interests or concerns.
- Understands the changing part of business in the public eye.
- Aligns social and ecological destinations with money-related objectives and explains the business benefits for seeking after social and natural goals.
- Designs and executes procedures connected to the association's center business methodology that enhance the association's social and ecological execution, with long haul dreams, medium-term objectives, and here and now designs and strategies.
- Applies sustainability to their part, capacity, and everyday choices and components sustainability learning into key, money related and operational basic leadership.
- Implements sustainability standards into authoritative capacities, approaches, and objectives.
- Uses situation arranging, back-throwing, and problem area and materiality investigation devices.

20.6.4.2. Progressed

Social reason and installation learning.

- Aware of how the organization makes, upgrades or decimates normal, social, relationship and human capital and its significance to business and societal achievement.

20.6.5. Active Values What Is It?

This portrays self and entire framework learning and mindfulness. It envelops the capacity to create and seek after higher reason inside self, groups, and business; to rehearse care, and to cultivate and empower individual and hierarchical change.

- **For what reason does it make a difference?:** High passionate knowledge cultivates strength, responsibility, trust, and correspondence between colleagues and outside partners, every one of whom are fundamental to empower business, societal, and transformational change. Similarly, dynamic esteems are expected to manufacture versatile abilities and strong businesses, enhance cooperation and coordinated effort and to surface, acknowledge and receive advancement. Care and pondering practices help change existing mental models and upgrade center and responsibility amid testing, vague and quickly evolving circumstances. An empathic and merciful outlook can create item and administration development lined up with center human esteems that are pertinent to clients' authentic needs and interests. Arrangement of profound individual esteems with business, society and the planet's needs can diminish pressure, enhance profitability and make opportunity and success.
- **What abilities does it expand upon?**: Decent variety, Integrity and Self-Management.

26.6.5.1. Establishment

Individual mindfulness and qualities arrangement:

- Regularly self-reflects structuredly; perceives and beats oblivious presumptions, blindsides, constraints and individual protection from change.
- Actively encounters and watches an idea or an inclination (particularly those that are negative and constraining), supposition, judgment, propensity, feeling or sensation, for example, fear, perplexity, strife or want, and after that forgoes quickly responding or reacting to the circumstance.
- Complements balanced idea with instinct, goals, fears, qualities, recollections and a feeling based approach.
- Understands possess feelings and needs and those of others (enthusiastic knowledge) and can see their own particular place in, and effect on, a circumstance (reflexiveness).

Exhibit humbleness – can state, "I don't have the foggiest idea" and request help; is guided by modesty and a feeling of appreciation and sympathy.

- Open to gaining from others and can let the unforeseen (and frequently exceptionally imaginative) develop in the midst of direness and time weights; open to thoughts that appear to be remote and notwithstanding undermining.
- Is a positive esteems good example inside and outside the association and show adherence to an arrangement of qualities in great and awful circumstances; advocates for dependable business administration inside and remotely.
- Aligns profound worry for social-biological trustworthiness to business parts and choices and is guided by a pledge to genuineness, truth, and straightforwardness in choices, systems, and plans.
- Transcends self-intrigue and concentrates on the group or more prominent great.
- Involved in businesses and occasions progressing biological and social sustainability.

26.6.5.2. Progressed

Individual change and higher reason:

- Is completely mindful and show at the time and consequently watches their considerations and sentiments and how they are relating, detecting, and judging through purposeful perceptions of different measurements of their self (counting the physical, mental, shadow, enthusiastic and otherworldly domains).
- Demonstrates genuine acknowledgment and graciousness toward every one of the measurements of themselves as well as other people, paying little heed to situation; reflects upon others without judgment.
- Possesses an attitude which sees the world, other individuals and all circumstances as a component of an entire in which everything is associated and settles on decisions and seeks after activities in light of the conviction that everything is between associated – moving past mindfulness to extraordinary mindfulness.
- Seeks individual change and seeks after formative encounters

which challenge their reality sees.

- Engages in transformational change activities that expand upon a profound association with themselves, the general population they work with, the planet and the more extensive world.
- Pursues work for a reason more noteworthy than themselves and sees work and business as a coordinated method to help change society; shows an inheritance outlook to tackle business to enhance prospects for present and who and what is to come.
- Creates a higher reason for themselves and adjusts their true nature to activities inside and outside business and, if what they are doing does not line up with reason, modify and upgrade it until the point when it does.
- Supports groups to take advantage of their instinct and develop aggregate knowledge; makes and catalyzes higher reason for others, drawing upon this for motivation and understanding.
- Fosters intelligent and generative discussions through the formation of precisely organized procedures of examination and reflection; makes opportunities for individuals and groups to address, investigate and make importance of the qualities and presumptions that advise the basic leadership process.
- Cultivates their own particular mindfulness and extends their reasoning through pensive and physical practices, for example, reflection, yoga, journaling, nature strolls, instructing, playing a melodic instrument, workmanship, judo, withdraws, peer coaching and groups of training.

Rehearsed together, these five skills will fabricate solid pioneers, groups, businesses, and social orders and position the business for progress over the medium to long haul. Businesses that enlist, select, create, boost and convey ability with these attributes won't just receive business rewards; they additionally make sound working environments, resolve individual business strains and acknowledge more amicable groups and social orders inside the biological limit of the planet.

20.7. RELATED COMPETENCIES

The five abilities distinguished through this exploration are exceedingly interlaced. One arrangement of practices fortifies another set, making an upright circle whereby one competency empowers and upgrades the others in a commonly gainful manner. This blend shows the basic idea of these five limits and their reliance as outlined beneath:

Systems thinking -> reveals more instrumental relationships -> opportunities for strategic collaboration -> social innovation which is enhanced by -> systems thinking and seeing a system from different viewpoints -> sustainability knowledge helps prioritize strategic investments and embed sustainability -> values-based self and system awareness -> fosters open-mindedness and enhances ability to collaborate -> social innovation, etc.

20.8. SUSTAINABILITY COMPETENCIES

Creativity and development are acknowledged through sense-production – the capacity to interface things to make new thoughts. To encourage social development it is basic to team up with outer accomplices and to see a framework in an unexpected way. It is critical to know about, and defeat, existing mental models in view of a prior time and experience. Too, sustainability learning advises the key concentration and catalyzes hierarchical change.

This association recommends that pioneer competency models ought to mirror every one of the five credits to guarantee an adjusted and compelling outcome.

20.9. ADVANCE TOWARD EMBEDDING SUSTAINABILITY

As businesses advance with sustainability and CSR, they ceaselessly insert sustainability advance into their human asset frameworks, corporate reason and plans of action. The diagram beneath demonstrates a run of the mill development way businesses seek after. It is a device that administrators, directors, and sheets can use to stock their advance (Table 20.1).

Table 20.1: Sustainability Embedment Path

Basic Focus	Compliance & Transaction Focus	Strategy Focus	Knowledge & Competency Focus	Business Model Focus
Define sustainability for the organization Ensure business case is understood Adopt a sustainability policy and vision	Embed in management system Embed in policies, procedures, employee code of conduct, basic communications	Embed in KPIs, corporate strategy, business and performance plans, bonusable goals	Embed in conventional, functional, core training Embed in competency model, leadership development and employee life cycle	Embed as core purpose Embed in corporate mission and vision

Source: ISO (2017).

20.10. WHAT'S NEXT?

These capabilities don't really characterize what makes individuals and firms fruitful today. They expect the ability ventures organizations should make to encourage business and social accomplishment throughout the following 10 to 20 years. They perceive that what made pioneers fruitful in the past may not prepare them to be viable later on. These five abilities are regarded the best ability speculations organizations should make to encourage both business and societal achievement in the decades to come.

While change begins at the best and requires vision, everybody has a part to play: sheets, officials, sustainability chiefs and HR experts are urged to fuse them into their competency models and utilize them to choose, create, send, remunerate, perceive and advance pioneers. By utilizing these five sustainability capabilities to characterize and drive superior, organizations will have the capacity to assemble an administration establishment to outline an effective course to what's to come.

Proficient affiliations, administration training, and business colleges ought to think about their part in outfitting future pioneers with these capabilities. HR, ability and learning, and improvement experts can recognize holes in their present way to deal with administration advancement and fabricate these authority qualities to empower future practical and business achievement. Businesses can utilize these skills to improve the ability pipeline and build up the up and coming age of pioneers and the authoritative abilities to control enterprises toward a sustainable future for all.

20.11. EXAMPLE OF SUSTAINABILITY PROFILE

20.11.1. Sustainability and Sustainable Development Expert

20.11.1.1. Responsibilities

The following responsibilities apply to project-wide activities as well as to specific pillars of the project if it were necessary:

- Project implementation and activities:
  - Coordination and management support on specific areas of the project, namely:

1. An Action Plan Framework for the Sustainability Strategy Initiative.
2. A Monitoring and Evaluation Framework for the Sustainability Strategy Initiative, including if necessary individual frameworks for the core components and key performance and reporting indicators.
3. A clear methodology for how the Sustainability Strategy Initiative will be monitored over time, with a reporting timeline to include the final evaluation.
4. A mid-term review of the Sustainability Strategy Initiative including a brief report on the preliminary results of the mid-term review, with a view to validating preliminary findings and testing tentative conclusions and/or recommendations during the final evaluation.
5. A final evaluation of the Sustainability Strategy Initiative, including a comprehensive report on the results and recommendations for all stakeholders. The report will be clear and simply written.

  - Assist with relationship management:

1. Designing a Stakeholder Engagement framework and procedure to ensure long- term relationships are forged and the needs of all involved taken into consideration.
2. Implement a Materiality Analysis framework to establish priority areas to be addressed throughout the project.

- Communications:
  - Assist with the development and implementation of marketing and communications activities, incl. Social media, newsletters, website, articles, and events for it

to be aligned with the Sustainability component of the project.
  - Putting together materials containing resources, trends and insights.
  - Support with opportunities to feature the project and team. This work can include research, submission of pitches for conferences, articles, etc.
  - Support and coordinate internal communication systems.
- Partnerships and Awareness:
  - Assist with the setting up and development of partnerships.
  - Support the development of broader awareness about the Project.
  - Use an entrepreneurial approach, working with other members of the team, to identify funding opportunities in different areas of the project.

20.11.1.2. Availability

The Responsibilities described will be performed on an online basis, as I will be traveling and will be available to go onsite as per September 2018.

20.11.1.3. Language Capacity

- The work will be delivered in English.
- Capacity to discuss and have meetings in French, Spanish & Portuguese if required.

20.12. CEO COMPETENCES

20.12.1. Introduction

The present corporate sheets have a major duty. With 16% CEO turnover and retirements a best driver of CEO renewal, their undertaking is to enlist the pioneers of tomorrow. These new pioneers will confront sensational change. They should explore the dangers and chances of fast population development and every one of that accompanies it – on the grounds that by the year 2050, our planet will have 9 billion individuals.

As of now investors, customers, employees, governments and host groups request that organizations successfully deal with their social and natural effects and help diagram a practical course for society. Deliberately shrewd organizations will upgrade their CSR and sustainability execution in the event that they wish to succeed and flourish in the twenty-first century. To steward an organization through this changing landscape – and profit by the business advantages of doing as such – organizations will require CSR-skilled pioneers in charge.

As this is a current business wonder, CEO progression arranging and enrollment programs regularly neglect to consider CSR qualities. To create and draw in pioneers with CSR esteems and capacities, board HR and enrollment advisory groups should refresh CEO position profiles and progression designs with new CSR criteria. In any case, while an ever-increasing number of sheets understand the significance of viable CSR initiative, they do not have the instruments and direction on the qualities required by this new organization official.

This guide tends to the holistic view. It assembles worldwide bits of knowledge into the best qualities CEOs and their successors should secure and make investor esteem through the quest for sustainability and CSR.

This guide furnishes sheets of executives with:

- The method of reasoning for including CSR and sustainability contemplations in CEO progression arranging and enrollment; and
- Recommended CSR criteria for incorporation in the determination models for the President and CEO. The rules depend on a writing audit of pioneer CSR qualities (recorded in the reference section) and an output of driving organization practices here.

The proposed criteria are intended to supplement – not dislodge – conventional business enlistment necessities. Indeed, some of the criteria expand and improve customary qualities. A perfect CSR position profile would incorporate at least one of these CSR criteria. At times existing criteria can be revised to factor in the significant CSR segment, while in different cases the CSR criteria may bring about new competency necessities. Emphatically dedicated CSR organizations may wish to incorporate them all. The primary segment gives six reasons why CSR ought to be joined into CEO choice models. The second segment points of interest the accompanying six CSR choice criteria for CEO progression arranging and enrollment:

- "Values" good example

- Externally mindful
- CSR strategist and change administrator
- Collaborates with partners
- Catalyst and backer
- Develops dependable pioneers

Forward-looking sheets and those engaged with CEO determination are urged to consolidate the CSR criteria most important to their plans of action and esteem insurance and creation procedures. Future-sealed businesses will require CEOs with the vision, duty, and know-how to deal with the dangers and grab the opportunities introduced by the changing business setting.

Here are six reasons why your board and HR advisory group ought to be engaged with refreshing the CEO progression arranging and determination criteria.

20.12.2. Six Motivations to Add CSR Criteria to CEO Progression Arranging and Choice

- **Successful CSR Execution:** Research uncovers that CEO administration is a basic achievement factor for successful CSR usage. Since CEOs set the "tone at the best" they have to show an individual sense of duty regarding, and faith in, the significance of CSR, so different pioneers and all staff can turn out to be completely engaged with CSR execution.

Keeping in mind the end goal to encourage CSR purchase in and take-up, CEOs need to reliably strengthen the organization's CSR duties and bearing through their own behavior and desires of others. Chiefs who just play lip administration to CSR will breed a culture of negativity and could cause a bargain of the organization's center esteems.

- **Brand and Notoriety Administration:** Since society anticipates that business will show great CSR administration, it is essential the CEO comprehends and models CSR states of mind, practices, and abilities that are lined up with partner esteems. Disregarding these authority limits may endanger the organization's image and notoriety. Research shows that numerous supervisors (43%) have been requested to accomplish something against their association's esteems. To keep this circumstance, it is vital that the CEO models for different pioneers and staff how to settle on choices that consider.

- To be sure scholastic research into the 2008 worldwide budgetary emergency uncovered that administration and administration character disappointments added to the unreasonable develop of use in money related markets and the ensuing meltdown. These character imperfections included carelessness prompting unnecessary hazard taking practices; absence of straightforwardness and now and again, an absence of uprightness; heedlessness to basic issues; absence of regard for people; and unreliability toward investors and the social orders inside which these businesses worked. (Accentuation included.)

Chiefs with solid CSR esteems, arrangement and limit are less inclined to have these character qualities and more averse to put their businesses and society in danger.

- **Employee Fascination, Maintenance, and Engagement:** In a tight work showcase a solid CSR-adjusted pioneer can better draw in and hold top staff. Research uncovers that CSR is additionally one of the key factors in worker engagement. CSR can rouse worker devotion and profitability where it exists, and can be a best danger to employee fulfillment where it isn't bolstered. Workforce soundness can be upgraded if the business shows a solid CSR responsibility in both word and deed. Chiefs need to "walk the discussion" to set a good example for staff and conquer inner skepticism or uncertainty, particularly for those staff whose preparation and foundation make it troublesome for them to value the esteem and significance of CSR.
- **Changing CSR Prerequisites:** Presidents need to comprehend that CSR is never again the consistence, magnanimous and advertising exercise it used to be. It is currently an expert business teach and board concern. More noteworthy desires, more open examination and expanded weights elevate concentrate on how organizations profit, the effect they have on the groups they work in, and the significance of making long haul shared esteem.

There are various CSR dangers that undermine business practicality, as recognized by the World Economic Forum. Later on, organizations should be stronger in an unpredictable, dubious, mind-boggling and questionable world. In spite of the fact that not really a CSR master, a keen CEO needs to comprehend CSR dangers and regard their importance to the organization, its market, its groups and particularly its clients. A CSR focal point on business can encourage advancement and opportunity creation, making CEO authority and support for CSR development a critical quality.

Today a CSR-proficient CEO needs to comprehend, oversee and advance the material social, natural and monetary dangers and chances of their business and those of its providers and clients. To guarantee their business has the limit and know-how to contend in this new working condition, the CEO needs a la mode thankfulness for the pertinence of CSR to business achievement.

- **President as Open Face:** Progressively the general population anticipates that organizations will lead on issues identified with thriving and versatility in the groups and markets in which they work. As organizations turn out to be all the more unmistakably connected with, the CEO likely will turn out to be more engaged with the general population exchange on CSR matters that influence its working condition. Having a CEO pioneer who can go about as a viable employee of the organization's social image will both improve its business and social objectives.
- **Better Choices:** Chiefs who have CSR capabilities can settle on and cultivate more compelling basic leadership inside the association. By applying a CSR point of view, their choices and those they bolster are more receptive, all-encompassing and long haul, profiting the supported accomplishment of the organization. Truth is told; CSR capabilities improve and fortify conventional business keenness and other alluring initiative characteristics, for example, a long haul see all-encompassing basic leadership, a hazard administration point of view and successful individuals' authority. Chiefs with solid CSR capacities are more viable pioneers in general.

These are six reasons a directorate may wish to incorporate CSR criteria in their CEO progression arranging and enlistment endeavors. Practically speaking, there are three CSR administration styles: CEOs who straightforwardly expel CSR; the individuals who express a nonexclusive responsibility, yet there is a hole between their words and activities; and CEOs who possess and are exceptionally engaged with CSR. The CSR criteria recorded in the following segment are intended to enable sheets to enroll CSR equipped pioneers – the individuals who claim and are engaged with CSR. They are proposed for thought by sheets, their HR advisory groups and the individuals who exhort them.

20.13. CSR LEADERSHIP CRITERIA FOR CEO SELECTION STANDARDS

The accompanying six CSR authority criteria can be utilized as a part of CEO progression arranging, initiative improvement, evaluation, and enrollment. The criteria can be utilized to educate CSR preparing for developing pioneers, inner CEO successors and as a focal point while picking another CEO.

The CSR authority criteria ought to be consolidated alongside exemplary and ordinary administration capabilities, a large number of which have a CSR or qualities based measurement, for example, morals and uprightness.

The accompanying criteria are drawn from worldwide research and a best-rehearse check. They include:

- "Values" good example;
- Externally mindful;
- CSR strategist and change administrator;
- Collaborates with partners;
- Catalyst and backer;
- Develops mindful pioneers.

A great part of the accompanying substance is verbatim from the worldwide research. CSR, values-based, dependable business and sustainability terms are utilized reciprocally. The CSR administration qualities incorporate a blend of learning, ability, practices, and states of mind.

Two of the criteria interface with regular or built up criteria, while the other four are new criteria for thought. Sheets and others engaged with CEO determination may wish to survey the accompanying criteria and select those (or parcels thereof), which are pertinent to their corporate setting for consideration in CEO enrollment and progression arranging forms. Businesses would thus be able to utilize the criteria to build up a tweaked CSR initiative approach.

20.13.1. Administration Development

The six CEO CSR Leadership Criteria can be utilized for senior administration ability enlistment, choice, improvement, maintenance, and examination and progression administration.

20.13.2. President CSR Leadership Defined

The CSR-able CEO makes an unmistakable vision in which individual esteems and business practices are firmly adjusted. The vision is conveyed reliably through their activities and choices and in their desires of others. They fabricate viable and incorporated administration that diagrams a course for business and societal achievement and those backings the top managerial staff to improve its sustainability limit and oversight. They support customary official discussions that distinguish and address the opportunities and dangers postured by worldwide sustainability patterns. They build up frameworks that reward sustainability advancement and execution. They guarantee credible engagement with partners, understanding that the result could impact the association's course. They champion sustainability to governments, industry pioneers and the speculation group. They understand this is a voyage that requires genuineness and fearlessness.

With bits of knowledge from business in the community.

1. **"Qualities" Good Example (New Criteria):** Is a qualities based pioneer. Individual esteems are lined up with hierarchical esteems. An extension esteems/esteem proclamations to real business practices and practices. Shows others how it's done in exhibiting the association's esteems and culture and is energetic about consolidating business and societal achievement. Works as per the most astounding social, natural and moral models.
    a. Acts as a good example inside and outside the organization. Models the way for other workers by demonstrating how they show the organization esteems in their decisions and practices. At the point when vital choices are made, they indicate how they mirror the association's esteems and impart this to staff. Examines with staff difficulties they look in maintaining the business and supports open, two-path correspondence on issues that emerge.
    b. Recognizes they give a noteworthy shadow a role as a pioneer and know about the effect of their shadow. In this manner, they both expressive a convincing and intelligent method of reasoning for the conduct they need to see, and exhibit it in what they demonstrate enthusiasm for, and how they invest their energy inside and outside work.

c. Is clear regarding why capable conduct matters to them, the business and to society. Knows that while discussing capable business is simple, openly conferring and finishing as a CSR pioneer requires strength.

d. Is reason-driven and ready to electrify the association around a typical reason.

2. Externally Mindful (Associates with Ordinary Criteria)

a. Demonstrates a strong attention to the mind-boggling sustainability issues that influence the business condition and sees the association between outer patterns and the suggestions for the center business. Comprehends and factors in worldwide social and condition inclines in business arranging.

b. Scans the skyline a long ways past the organization and industry to understand what is occurring in business and society everywhere and recognize new opportunities. Invests energy with individuals inside and outside the organization gathering data from both formal and casual channels (counting sites and other online networking), and from systems of "various reasoning "individuals.

c. Explores "jostling notes" (flags that are awkward, genuinely unclear toward the begin, however, could be vital) and is set up to think the unfathomable, regardless of whether the suggestions may be distressing. Evaluates low likelihood/high effect hazards and entwined dangers that could endanger the organization's future. Considers "feeble signs" notwithstanding when their effect won't not be promptly self-evident.

d. Takes an all-encompassing way to deal with arranging an investigation that considers the way a framework's constituent parts interrelate, how frameworks work after some time and inside the setting of bigger frameworks – where the framework could be the organization's esteem chain, who and what is to come, partners, ecological conditions, geopolitical elements and different contemplations. For instance, they know about the association's long haul relationship with the nearby/have group. They comprehend the organization's neighborhood impression and heritage over the long haul and the part the organization can play to add to nearby thriving.

e. Positions the business for short-and long haul accomplishment by driving the organization to adjust to – and impact – basic social, monetary and natural patterns.

3. **CSR Strategist and Change Administrator (Associates with Customary Criteria)**

a. Leads the advancement of, and imparts, a convincing future (vision) for the business – one that mirrors the association's social obligations, makes an incentive for society, and perceives the shifting goals and desires of partners and searches out the business openings that sustainability offers. Stewards the improvement of a plan of action and system that adjusts social and natural targets to monetary objectives.

b. Understands how to implant CSR into business choices. Sees the master plan, takes the long view and comprehends both business and societal ramifications of choices. Parities budgetary, operational and here and now business weights with the long haul sustainability of the business and society. Chief's CSR esteems are maintained in both great and awful circumstances and shape business choices at all levels (particularly troublesome choices like redundancies).

c. Understands and can explain the business case for seeking after social and natural targets. Can clarify how CSR converts into esteem creation particularly for the business. Distinguishes the cooperative energy between business movement and sustainability.

d. Leads complex social change inside their businesses to implant sustainability. Encourages authoritative limit with respect to change and advancement to make feasible plans of action and answers for what's to come. Qualities assorted variety of thought, foundation, point of view and culture. Examinations with new thoughts and arrangements and invites radical discussions. Leads the advancement of practical items and administrations with a view to having a beneficial outcome all through the esteem chain.

4. **Collaborates with Partners (New Criteria)**

a. Can recognize key partners that have effect on the association. Sees how the association impacts key partners emphatically and contrarily. Shows an enthusiasm for and information of developing partner slant and desires and can react insightfully and consciously to contending partner interests.

b. Can lead change past business limits. Includes partners to help characterize and make esteem. Gets by and by included and draws in with communities, NGOs, scholastics and other people who challenge with various perspectives and can be a wellspring of inventiveness and advancement. Has a reputation of delivering budgetary outcomes with due thought for the interests and worries of partners.

c. Listens to individuals who question or don't concur with them or the organization's bearing and can extricate experiences from such dispute. Creates and keeps up assorted businesses with partners, currently inviting alternate points of view and esteeming distinction, producing inventiveness and lessening bunch think.

d. Understands and qualities the basic of teaming up with outside partners to address challenges that require aggregate fundamental arrangements. Guarantees the association has interior limit with regards to successful partner joint effort on CSR issues.

e. Builds activity arranged, commonly gainful businesses and coordinated efforts incorporating with offbeat partners. Can work straightforwardly and successfully in organizations and joint efforts. Makes progress in building outer impact and seeks after industry and partner administration and connections to make an incentive for the association and society.

5. **Catalyst and Supporter (New Criteria):** Is a supporter for capable business authority. Perceives that business must be a basic piece of the answer for neighborhood, national and worldwide social and ecological difficulties. Understands that gaining ground on issues like neediness, regular asset utilize, joblessness, training and building flexible and prosperous groups requires joint effort among business pioneers. Adds to open level

headed discussion on these issues with an educated perspective.

a. Promotes sustainability to governments, clients, providers, and business accomplices. Offers a dream on the social reason and obligations of business and on the business profits by this approach.
b. Champions sustainability to workers and leads the association to be an impetus for a feasible society.
c. Advocates with government and industry relationship on the requirement for open arrangement that empowers a sustainable society. Plays an influential position in industry and expert relationship in the advancement of sectoral and expert CSR gauges.
d. Challenges short-termism freely and secretly. Imparts long haul sustainability and long haul esteem creation to investors, financiers, and safety net providers and at AGMs.

6. **Develops Mindful Pioneers (New Criteria):** Actively looks for individuals with a CSR and qualities based authority viewpoint while enrolling into the association and makes this a need in administration advancement programs.
   a. Mentors and creates future pioneers who comprehend more extensive societal issues, their part and the part of business.
   b. Uses reward-and-acknowledgment frameworks and correspondences to reinforce positive practices and results. This implies coordinating CSR and value-based business practices into group targets, individual execution surveys and KPIs. Conveys why these qualities are vital.
   c. Creates the conditions to energize administration, social enterprise endeavor and better approaches for speculation to rise.
   d. Provides opportunities for supervisors and future pioneers to increase social encounters and bits of knowledge by working with a societal association or activity. Perceives and gains by the esteem that social encounters and on-the-ground network convey to the business.

20.14. CONCLUSION

Sheets of executives are progressively embracing CSR commands and getting to be plainly engaged with CSR oversight and stewardship. When they comprehend the hugeness of CSR to the company's future execution and CSR ends up plainly installed in the corporate methodology, they will need to guarantee their CEO has the skills and qualities arrangement to empower powerful execution. This makes it basic those sheets, and the individuals who prompt them, have the bits of knowledge they have to incorporate CSR contemplations in CEO progression arranging and inquiry. Various reports have been created regarding this matter and driving organizations and administrators are demonstrating the way.

This is an emerging board governance topic that will continue to grow in importance as CSR becomes an increasingly significant driver of value creation for organizations. With CEO turnover due to retirements, this is the time for boards to enhance their knowledge of these critical competencies and embed them in their recruitment criteria. Well-governed boards now have a tool to help them with this essential task.

CHAPTER 21

Chain Liability: Responsibility, Reputation, and Accountability

When it turns out to be openly realized that items are related with providers that participate in unsustainable practices, customers challenge, as Nestlé, Zara, and Kimberly Clark, among others, have learned. The phenomenon by which shoppers consider firms in charge of the unsustainable conduct of their upstream accomplices recommends the idea of "chain risk." This investigation expects to produce bits of knowledge into the predecessors and outcomes of such customer duty attributions. Utilizing information from four vignette-based survey tests, the creators find that the chain risk impact increments if an ecological corruption episode (1) comes about because of provider conduct as opposed to constrain majeure, (2) comes about because of an organization choice as opposed to the choice of an individual employee, and (3) is more extreme. Duty attributions don't contrast with changing authoritative separation from the provider, firm size, vital significance of the provided item, or the presence of ecological administration frameworks. The chain obligation impact likewise makes solid dangers for the central firm; higher duty attributions increment buyers' outrage and inclination to blacklist. In this manner, firms should work to guarantee sustainable conduct all through the production network, to shield them from chain risk.

21.1. INTRODUCTION

An allegation posted on Nestlé's Facebook site in April 2010, "You have annihilated a large number of hectares of rainforest. I will never again purchase any Nestlé items," is illustrative of many comparative posts by angered buyers. These challenges turned out to be particularly unmistakable when open reports uncovered that the palm oil contained in a considerable lot of Nestlé's items was gotten through unsustainable practices: Sinar Mas, one of Nestlé's providers, chopped down tremendous swaths of rainforest to build the measure of land accessible for its palm oil creation. Purchasers held Nestlé responsible, and this case isn't exceptional: Kimberly Clark was despised when its tissue providers collected chronicled forests, and Zara was faulted when subcontractors utilized lethal chemicals to color clothing.

Obviously, purchasers don't separate between individuals from the inventory network with regards to unsustainable conduct. Rather, they consider the central firm in charge of everything that happens in the production network, which makes a "chain obligation impact." Such unsustainable practices may involve monetary, social, or potentially natural measurements (e.g., Elkington, 2004), however for this examination, we principally concentrate on the ecological measurement, which eventually influences human prosperity (Gao and Bansal, 2013). In this unique circumstance, unsustainable provider conduct alludes to provider practices that diminish biological capital and damage the earth (Andersson and Lindroth, 2001); it contrasts from poor execution or provider wrongdoing, which rather suggest general aberrance from fundamental store network administration goals (e.g., cost, quality, benefit, adaptability). Unsustainable provider conduct rather expands the danger of unsustainable occurrences, either promptly (e.g., tainting of water supply) or after some time (e.g., environmental change because of ozone-depleting substance emanations). For example, Chinese directions preclude the arrival of harmful effluents into conduits. These directions are every now and again disregarded by suppliers to the dress (e.g., Zara) and different businesses who regularly furtively discharge lethal chemicals into water frameworks (Greenpeace, 2009). These unreliable demonstrations speak to unsustainable conduct. They prompt serious negative results for the earth and nearby tenants. One particular poison caused the passing of 8000 individuals in four noteworthy Chinese urban communities which constitutes an unsustainable episode (Economy, 2013). That is, we utilize "incimark" to allude to a negative result of unsustainable conduct. We additionally take note of that numerous customers express their eagerness to blacklist firms related with an unsustainable occurrence, with considerable monetary hazard (Klein and Dawar, 2004; Kovács, 2008).

In spite of such effects and their pertinence, despite everything we do not have a reasonable comprehension of the hidden causal deductions by which consumer's property duty regarding unsustainable provider conduct to a central firm (i.e., chain obligation). Writing on store network sustainability offers essential commitments (Carter and Rogers, 2008;

Seuring and Mueller, 2008) yet additionally endures some persevering holes. For instance, store network writing predominantly concentrates on the advantages of sustainability, as opposed to the dangers related with insufficient endeavors (e.g., Carter, 2000; Jayaraman et al., 2005; Melnyk et al., 2003). In examining store network structures, earlier research tends to coordinate provider connections, with just minimal attention committed to multitier supply chains (Kovács, 2008; Mena et al., 2013). At long last, existing investigations diagram approaches to guarantee provider consistence with desires and principles (Delmas and Montiel, 2009; Jiang, 2009), to urge providers to wind up plainly more manage capable (Gavronski et al., 2011; Klassen and Vereecke, 2012; Vachon and Klassen, 2008), and to decide the impacts of such activities on execution (Bai and Sarkis, 2010; Carter and Jennings, 2004; Zhu and Sarkis, 2004), however they overlook the impact of the absence of such exercises.

Operations and store network writing normally pays less attention to buyers' discernments; however, a few examinations incorporate this point of view, concentrating on the results of obligation attributions as opposed to their determinants. We hence discover examinations of dangers to customer security radiating from provider wrong-doing and their suggestions (e.g., Hora et al., 2011; Marucheck et al., 2011). Promoting ponders likewise endeavor to explain the determinants of obligation attribution by noticing how an absence of sustainability can be show in flawed or risky (Gao et al., 2012; Klein and Dawar, 2004; Laufer et al., 2005), instead of in the creation procedure. Regardless of these significant contributions, to the best of our insight, no examinations incorporate natural measurements of sustainability worries in multitier supply chains, in which customers' desires decide their attributions and characterize their responses.

As needs be, this examination tries to make three commitments. Initially, we intend to decide whether customers consider central firms in charge of the environmentally unsustainable conduct of their providers in upstream, multi-tier supply chains, or if this impact is episodic. With a vignette-based examination, we acquire prove that shoppers accuse the central firm for unsustainable provider conduct and along these lines clear up buyers' desires of ecological principles in a multitier store network.

Second, we research relevant variables that may increment or diminishing such duty attributions, identified with the occurrence, the association with the provider, and the structure of the store network. By considering the relative influences of eight elements, we offer basic bits of knowledge for inventory network leaders, who must organize and advance focused on activities and engagement. Third, we shed light on purchasers' enthusiastic and behavioral responses to duty attributions, to offer bits of knowledge into the potential market dangers related with an unsustainable episode. That is, we inspect unsustainability in supply chains by recognizing buyer discernments and responses as a beginning stage for a meaning of proper production network procedures.

In the following area, we present attribution hypothesis as our hypothetical premise and characterize the key builds that prompt our speculation improvement.

21.2. ATTRIBUTION HYPOTHESIS

Attribution hypothesis (AT) sees people as normal information processors who attempt to understand the world by ascribing causes to episodes. Looked with an episode, a man tries to understand what caused it and figure out who is mindful (Fincham and Jaspers, 1980; Hamilton, 1978). As indicated by AT, obligation attributions change on a few measurements. The locus of causality differentiates situational components and capacities; controllability alludes to the level of volitional control a wrongdoer has over the result, speaking to the level of direct impact on the reason. Security is how much the reason stays consistent after some time and shows what's in store from the guilty party in future, in comparable circumstances (Weiner, 1986, 1995). Moreover, an expansion of AT incorporates occurrence seriousness, with the end goal that individuals likely trait greater duty regarding more extreme episodes (e.g., Tennen and Affleck, 1990). Hence, customers ought to relegate greater obligation if a reason for an episode is (an) inward, (b) controllable, (c) stable, and (d) more extreme.

Obligation attributions are innately fixing to enthusiastic and behavioral responses (Weiner, 1986, 1995). For instance, neediness attributions that concentration past the individual (e.g., ruined because of a cataclysmic event) have a tendency to incite sensitivity as an enthusiastic response and help as a behavioral response. Attributions that address the individual's own particular activities (e.g., betting) rather likely prompt negative feelings and behavioral responses (e.g., declining to help; Weiner, 1986, 1995).

In operations administration writing, AT principally has served to explain the social, behavioral, and enthusiastic parts of musical show basic leadership (Bendoly et al., 2006; Urda and Loch, 2013), including hold up observations and their results (Seawright and Sampson, 2007), the impacts of client businesses on purchasing practices in self-benefit circumstances (Li et al., 2013), the impact of reasonableness discernments on the level of participation for innovation enterprise (van Burg and van Oorschot, 2013), and the social, mechanical, or individual factors that prompt (un) safe work behaviors (Brown et al., 2000; DeJoy, 1994). Eckerd et al., (2013) use AT to see how ruptures of mental contracts influence function practices and attitudinal discernments, with regards to two inventory network accomplices. Conversely, most operations administration look into applies at the individual level. With this examination, we stretch out the use of AT to occurrences that happen in multitier supply chains, as saw by singular customers, to explore sustainability in supply chains.

Attribution hypothesis additionally can reveal insight into customer perceptions of corporate duty. Skarmeas and Leonidou (2013) find that attributions of prideful and partner-driven thought processes evoke purchaser incredulity; values-driven attributions hinder such distrust. Thus, suspicion brings down retailer value, diminishes protection from negative data about the retailer, and stimulates horrible informal. Buyers who rather see capable activities as earnest and in accordance with the general organization procedure remunerate organizations by picking their items over competitors' (Becker-Olsen et al., 2006; Ellen et al., 2006; Yoon et al., 2006. In this way, shoppers seem to remunerate practical (dependable) firm conduct; we likewise look to see how buyers rebuff unsustainable practices.

In created economies, customers push firms to screen and close holes in their sustainability execution crosswise over supply chains (Klassen and Vereecke, 2012; Parmigiani et al., 2011), considering them in charge of the conditions in which the items they offer were produced (Phillips and Caldwell, 2005). Obligation is unequivocally given when the firm can make autonomous decisions and control and impact results in the production network (New, 2004; Parmigiani et al., 2011). Such expert and capacity rely upon various elements, as do shoppers' duty attributions.

21.3. DISCUSSION AND IMPLICATIONS

Our discoveries bolster the presence of a chain risk impact in multitier supply chains: A central firm might be considered in charge of its own as well as its providers' unsustainable practices. Despite the fact that an episode coming about because of power majeure diminishes consumers' obligation attributions (H1), none of our theories identified with the controllability measurement of AT prompted altogether bring down duty attributions (H2–

H5). Along these lines, we demonstrate that chain risk holds on, regardless of whether the unsustainable conduct happens outside the central firm (H2), at an authoritatively far off provider (H3), at a substantial or little firm (H4), or in connection to pretty many imperative provided parts (H5). We offer two possible clarifications for these discoveries. Initially, shoppers may not completely fathom the troubles of overseeing global, multi-tier supply chains. To diminish chain risk, central firms should seriously think about illuminating buyers about the (restricted) potential outcomes for set up straightforwardness and control over complex chains. Then again, the showcasing writing recommends a more nuanced clarification: Gao et al., (2012) contend that shoppers confronting emergencies that influence different organizations receive routine and advantageous demeanors and practices, regardless of realizing that the issue requires broad, complex arrangements. The look for a "substitute" offers a basic, helpful method for dealing with stress. Correspondingly, on account of item hurt emergencies, negative overflow impacts spread all through the pushing family (Roehm and Tybout, 2006), which may likewise apply to "chain families." Consumers may perceive the troubles of guaranteeing sustainable provider conduct all through the chain yet at the same time think that it's more helpful to accuse a central firm.

A third arrangement of our speculations identified with the security of an unsustainable episode. In spite of the fact that we anticipated greater dependability perceptions when the episode was because of organization practices (H6) and no EMS existed (H7), we found that purchasers recognize just that worker conduct can never be totally managed by the business, so they characteristic less duty regarding an occurrence caused by a employee. They don't generally think about the company's push to build up an EMS. Maybe organizations that take part in sustainable inventory network administration make desires that no occurrences will happen, in light of the fact that generally, their exercises speak to "greenwashing" (Kim and Lyon, 2011).

Duty attributions by purchasers along these lines reflect factors over which a central firm viably has little control or impact (New, 2004; Parmigiani et al., 2011).

These discoveries are basic; eventually, shopper purchasing behavior decides the mischief to organizations. The more the central firm is considered in charge of unsustainable conduct in the inventory network, the more probable purchasers are to respond with negative emotion (H9) and conduct

(H10). Along these lines, our discoveries loan exact help to Frooman's (1999) partner impact show. On the off chance that a partner—the shopper in our examination—does not have any chance to impact an association specifically, it must apply a backhanded influence procedure. In this way, a purchaser may try to impact the central firm, which has impact over the objective unsustainable organization, to drive it to act economically. In the event that purchasers withhold their cash from the central firm, they might attempt to compel it to guarantee sustainability all through its inventory network.

Another outcome additionally merits talking about. Feelings don't generally intercede the connection between duty attributions and behavioral responses, conversely with the AT, which places that obligation attributions prompt negative feelings (outrage), which at that point deliver behavioral responses (boycotting). Nonetheless, in two of our four trials duty attributions specifically influenced boycotting; feelings were full middle people for pants (Dressup) and screens/shows (Visual life); however, reciprocal go-betweens for glues (Klebfix) and autos (Fastcar). This hypothetically intriguing outcome may suggest the substitutability or low significance of the last two items. In the event that elective items offer great substitutes (autos) or the item is insignificant (glues), a buyer may blacklist after an unsustainable episode, without creating negative feelings. Behavioral changes without going before feelings are sound just if the apparent exertion of changing items or suppliers, because of a blacklist, is low.

21.4. THEORETICAL IMPLICATIONS

Research on reasonable inventory network administration, for the most part, concentrates on the advantages for central firms, as far as expanded environmental, social, or monetary execution (Bai and Sarkis, 2010; Carter and Jennings, 2004; Zhu and Sarkis, 2004). Far less consideration has tended to the valuation of an absence of sustainability (Auger et al., 2008, 2010), despite the fact that negative occurrences have more noteworthy effects than positive ones (Sen and Battacharya, 2001). With this exploration, we particularly address the outcomes that emerge for central firms because of unsustainable provider conduct and exhibit that this area merits measure up to consideration.

Research ought to recognize that the idea of such negative episodes changes as far as territory and influenced substance. In musical shows administration investigates, negative episodes regularly show up under the umbrella subject of item hurt emergencies (Dawar and Pillutla, 2000). Marucheck et al., (2011) contend that real item absconds regularly result from upstream activities in worldwide creation chains. Moreover, customers don't blacklist just when their own wellbeing and prosperity is influenced; they likewise do as such in light of the fact that assembling forms initiate negative impacts somewhere else. Along these lines, multi-tier supply chains must offer item traceability, item lifecycle investigation, and review administration.

Operations administration writing, for the most part, recognizes outside partner weights as drivers of sustainability initiatives, coordinated toward the upstream esteem chain. We improve this stream by giving all the more fine-grained examinations of customer perspectives on unsustainable occurrences in supply chains. Central firms endure exposures to the chain risk impact, regardless of whether an episode happens at the site of a far off, upstream provider. Awaysheh and Klassen (2010) contend that as separation builds, firms confront greater many-sided quality and exchange costs in their cooperations with suppliers, driving them to utilize all the more observing and evaluating frameworks to relieve data and straightforwardness issues. We take note of the significance of growing such endeavors past level 1 provider; however checking frameworks coordinated in danger moderation constitute only an initial phase in meeting high buyer desires (Klassen and Vereecke, 2012). A sensible second step is upholding changes in provider behavior or enhanced provider natural execution. Just to the degree that checking prompts provider conduct redress would it be able to shield a central firm from chain risk.

At last, this exploration adds to AT writing; we stretch out its application to the interface between an individual purchaser and a system of businesses in multitier supply chains. Attribution hypothesis already has been utilized predominantly to clarify the causal derivations of a purchaser about someone else (e.g., Crossley, 2009; Fragale et al., 2009) or association (e.g., Taylor, 1994). Albeit got consistently from AT, our controls, particularly those related with the controllability measurement, were not affirmed. In an inter-organizational setting, the controllability measurement shows up excessively complex for singular buyers, making it impossible to genuine. Furthermore, AT recommends that passionate responses completely intercede the connection between obligation attributions and behavioral responses, yet we discover a few occurrences in which enthusiastic responses just in part intervene this connection. The immediate way from obligation attributions to behavioral responses along these lines ought to be considered and examined in uses of this hypothesis.

21.5. MANAGERIAL IMPLICATIONS

On account of an unsustainable episode in the inventory network, the market will rebuff the central firm for its providers' conduct. Approximately 75% of our respondents showed they would not purchase from the central firm in the wake of having found out around an unsustainable scratch in its store network. In this way, we give additional confirmation of the critical part that clients, as partners, serve in determining the selection of reasonable systems (Hillman and Keim, 2001). Chiefs must create abilities and capacities to comprehend the desires of partner gatherings, and clients specifically, for sustainable conduct (Harrison et al., 2010). This proposal additionally recommends the significance of fortifying the hierarchical connection between store network administration and showcasing.

Past creating abilities to comprehend the desires of the market, organizations should work to guarantee sustainable conduct all through the production network, to shield themselves from chain liability. We suggest all organizations, not simply vast firms, do as such. We additionally take note of that organizations can't concentrate just on their expansive, deliberately essential, level 1 providers; they should go past such dominating methodologies (Chatain, 2011; Surroca et al., 2010). In addition, firm discoveries that propose production network directors ought to grow more significant capacities for provider advancement, collaboration, and development (Gavronski et al., 2011; Klassen and Vereecke, 2012), at that point exchange those abilities crosswise over levels (Mena et al., 2013). For instance, IBM's Social and Environmental Management System, executed among its around 30,000 level 1 sup- pincers, causes them implement directions among their providers, that is, among the level 2 providers of IBM.

On the off chance that an occurrence happens, all around created correspondence techniques must be set up to impart data to customers. Nonetheless, as per our discoveries, the main factors that can relieve the weight of being considered responsible for unsustainable conduct by providers is if the occurrence is because of power majeure, is less extreme, or emerged in light of the fact that a employee acted freely, not in soundness with organization practices.

21.6. LIMITATIONS

The decisions we made for this examination suggest certain constraints. To begin with, we concentrated on ecological practices and occurrences in the production network, in light of the fact that natural debasement pretty much straightforwardly and instantly influences human prosperity. Our manipulation of episode seriousness consolidated social and natural measurements of sustainability, however, which may perplex our outcomes. Second, the items changed considerably over the four exploratory vignettes (glues, attire, autos, screens), ranging from utilitarian to hedonic, from easy to complex, and from chemicals to innovative items. These distinctions suggest more noteworthy generalizability of our discoveries; however, they additionally could impact buyer view of the production network. Third, our single-nation center brings up the issue of whether customers from different nations respond comparably when looked with unsustainable conduct in a store network.

German purchasers have notoriety for being extremely delicate to natural insurance, so they may have elevated standards about inventory network sustainability. Fourth, the utilization of vignettes just enabled us to evaluate shoppers' proposed responsibility attributions and enthusiastic and behavioral responses; we didn't straightforwardly quantify genuine purchaser responses to genuine unsustainable episodes. Shoppers' planned responses frequently contrast from their genuine practices, as exhibited in investigations of purchaser purchasing conduct for items with natural highlights (Auger and Devinney, 2007).

21.7. CONCLUSIONS AND FURTHER RESEARCH

The target of this examination has been to expand our comprehending of whether and when purchasers consider central firms in charge of the biologically unsustainable conduct of their providers in upstream supply chains, utilizing both the indicators and results of such obligation attributions. We placed that eight components derived from four AT measurements (i.e., locus of causality, controllability, soundness, and seriousness) would identify with duty attribution. Nonetheless, our four vignette- based investigations uncover that exclusive power majeure, a background marked by fruitful EMS implementation, and occurrence seriousness altogether influence the degree to which purchasers consider the central firm in charge of unsustainable behavior in the production network. No indicators identified with controllability were noteworthy; shoppers don't recognize confinements on the company's control inside the store network. Higher obligation attributions identify with more grounded sentiments of outrage and a higher penchant to blacklist, which features the monetary hazard to central firms. Consequently, this examination builds up the chain obligation impact: Consumers consider a central firm in charge of practices and occurrences upstream.

Extra research is justified to decide how inventory network directors can viably and proficiently broaden their compass in the store network, to build straightforwardness, information, and control. Analysts could investigate the ramifications of negative episodes for the monetary execution of the whole store network. Interestingly with AT, our intervention examination showed that customers blacklist a few organizations while never feeling negative feelings; chance administration looks into subsequently ought to investigate nearly which factors trigger a prompt discipline, free of negative feelings. We additionally suggest that blemish scientists ought to decide if the expenses of endeavors to avert unsustainable provider conduct can be exchanged to the market. That is, does sustainable inventory network administration increment shoppers' ability to pay? Concerning our finding that shoppers accuse the central firm, it is intriguing to discover in the event that they neglect to appreciate the central company's restricted control over level 2 and level 3 providers, or on the off chance that they couldn't care less and essentially need to rebuff the most available substance, in particular, the central firm. At last, we concentrated on ecological occurrences; however, it might be similarly enthusiasm to investigate the predecessors and results of obligation attributions for negative social episodes. Additionally inquire about along these lines could examine buyer observations and responses to socially unsustainable provider conduct.

CHAPTER 22

Communication and Marketing

22.1. INTRODUCTION

Regardless of where you are in your concede cycle, making and depending on a correspondence design grew particularly for your give site as a team with your accomplices and partners can be a basic segment of your drive's long haul achievement. A very much-considered arrangement can give a vital guide to your correspondence exercises. This can be especially valid for Safe Schools/Healthy Students (SS/HS) activities setting out on an arrangement for sustainability.

Making a correspondence anticipate sustainability presents you and your accomplices with a chance to recommit as a group to managing achievement. Your arrangement will encourage reveal approaches to extend existing businesses and grow new ones that hold the possibility to help support—or even extend—basic elements of your drive. Your arrangement will enable you to benefit as much as possible from your coalition's restricted time and assets. Having an arrangement set up can help lighten the pressure numerous grantees feel close to the finish of Federal financing—push that is regularly joined by the inquiry, "What on earth do we do now?"

Possibly you and your accomplices made a correspondence design at a very early stage in your concede cycle. We unquestionably trust so! In the event that you did, now is the ideal time to return to that procedure, since you'll be connecting with new and diverse gatherings of people to recount your story and create bolster for what's to come.

Or on the other hand, maybe you and your accomplices have been conveying all over to advance your drive—a public statement submitted to your neighborhood paper declaring your allow grant route back when, a leaflet that depicts your projects and administrations, a Web webpage that is refreshed sometimes. Conceivably you have felt as if there has never been sufficient time to make a bigger arrangement. We comprehend those limitations. Be that as it may, we would propose that the correspondence strategies you're sporadically utilizing could have a substantially more prominent effect if made inside the setting of a general procedure.

Maybe you believe it's past the point where it is possible to begin a correspondence design. We oppose this idea. Truth be told, the Communication and Social Marketing (CSM) Center trusts it's never past the point of no return—to ponder correspondence.

On the off chance that you've gone to a CSM Center preparing, you've been acquainted with our 8-Step Communication Planning Model. This archive comes back to that model and is intended to give you the establishment for leaving on that arranging procedure. We trust it will give you gratefulness for the potential benefit of arranging correspondence to help your sustainability, and we trust you and your accomplices will make the sort of plan that can really help your drive leave an enduring impression in your group.

Also, recall—specialized help (TA) for correspondence getting ready for sustainability is one of the many administrations offered by the CSM Center. Your specialized authority would be cheerful to work with you anytime to decide availability for correspondence arranging and help see you through the procedure.

22.2. THE 8-STEP COMMUNICATION PLANNING MODEL

The CSM Center uses the 8-Step Communication Planning Model while giving TA in correspondence arranging. This model gives a successful diagram to correspondence arranging, paying little mind to where you are in your allow cycle. It depends on the hypothesis and practice of social advertising, which adjusts business-showcasing systems to impact individuals to make suggested move or roll out positive improvements in their lives. The eight stages are:

- Start with an exhaustive appraisal of your present circumstance;
- Set quantifiable correspondence objectives and targets;
- Define your target groups;
- Develop and pretest your messages to those gatherings of people;

- Select the most ideal approaches to convey those messages;
- Make an activity get ready for correspondences;
- Develop and pretest any materials you make; and
- Finally, actualize your arrangement—assessing and adjusting as required.

Before Beginning a Communication Plan for Sustainability partnerships:

- Your first undertaking might be among the most troublesome. Your drive should distinguish those capacities reserved for sustainability. We ask you to incorporate your TA expert (TAS), evaluator, accomplices, and partners in this frequently difficult work. Intense choices may be made; however SS/HS grantees can profit by taking a gander at their information for direction: What's functioning admirably? What are the best needs?
- Frequently, destinations will concentrate on maintaining employments that have been subsidized by the concede, yet we empower you and your accomplices to recall that an objective of the SS/HS Initiative is to make enduring fundamental group bolstered change. This may imply that you and your accomplices should move the center from maintaining singular positions to managing those capacities that give the significant help your kids and families require.

1. **We're Ready To Plan—What's First?**

- We trust that the best correspondence designs are created by a group that stretches out past the neighborhood SS/HS venture chief and staff individuals to incorporate your evaluator, accomplices, and key partners. Your partners could comprise of group offices, business and religious accomplices, school pioneers, and region chairmen.
- Consider uniting those individuals who eagerly share your objective of making schools and group in which families are bolstered, and kids are sheltered to settle on sound decisions in all aspects of their lives. By welcoming partners into the correspondence arranging process, you will accomplish a level of purchase in for your sustainability endeavors that may expand your odds for progress.
- Would We Be Able to Do This All Alone?

A few grantees like to deal with this procedure in little additions with their accomplices. This prologue to the procedure is intended to get the arranging procedure off on the correct foot and keep you on track en route. Different grantees ask for help with encouraging the correspondence arranging process.

22.2.1. Stage 1: Assess Current Situation

Reason: To practically evaluate your circumstance and the difficulties you look as you push toward sustainability. To explain your victories to date, and in addition, you're apparent difficulties pushing ahead.

In this basic initial step, you and your accomplices altogether portray where you are at the present time. Why? Since you can't get ready for the future on the off chance that you aren't straightforward about where you are today. You should investigate the realities, and also the intangibles encompassing your drive and group.

a. **Initially, Ask the Right Questions:** Begin by asking the correct inquiries—the appropriate responses will give you a solid rude awakening and will incorporate both the hard facts and the intangibles encompassing your program. Give this rundown a chance to give a state of flight to your aggregate reasoning:

- Have you recognized the capacities you need to support?
- Are you serving the quantity of understudies and families you set out to serve?
- Are you adequately serving the assortment of societies in your group?
- Are your instructors and principals' content with your program? Is it accurate to say that they are "champions" yet?
- Is your director getting comes about? Have you been displaying to your school board regularly?
- Are you content with your result information? It is safe to say that they are successfully driving choices?
- Have you had any fruitful occasions around your drive? Great media scope?
- Are your organizations focused on managing this work?
- How does cash stream in your group? How is your opposition financed?
- Whom else do you require on board to manage your drive? Do you see your central goal aligning with different businesses in

your group? At that point, consider conducting a SWOT analysis An investigation of your qualities, shortcomings, openings, and dangers (SWOT) is a brilliant method to assess where your program is at the present time. It is an astounding structure for participating in exchange between accomplices, staff, and different partners.

- Strengths are inside characteristics of your program that can help accomplish your destinations.
- Weaknesses are inward traits of your program that can thwart your targets.
- Opportunities are outer conditions that can help accomplish your destinations.
- Threats are outside conditions that can frustrate the program's execution.

22.2.2. Stage 2: Set Communication Goals

Reason: To express what it is that you might want individuals to do that they're not as of now doing in ways that are sensible and quantifiable.

Before handling this progression, you and your accomplices should recognize those basic elements of your drive that you'll work to manage. At exactly that point will you be really prepared to set correspondence objectives that will serve your decisions.

What qualities make for a solid objective? For the most part, the best correspondence objectives meet these criteria:

- ***Prioritize:*** Your correspondence objectives will concentrate on those capacities that your organization has managed. Since it is for all intents and purposes difficult to maintain each program, administration, and position initially supported by your SS/HS give, you'll need to focus on your set up needs for sustainability. Endeavoring to achieve everything regularly implies we prevail at nothing.
- ***Are Reasonable:*** Do your objectives pass "the creative ability test"? At the end of the day, would you be able to envision contacting them? In the event that you can't imagine achievement, there's a decent shot your objectives are not reasonable. While surveying the feasibility of your proposed objectives, think about accessible assets and potential wellsprings of help. For instance, your school area might confront serious spending slices that make it improbable to expect preceded with locale subsi-

dizing for your prominent after-school programs. In any case, you might have the capacity to create businesses with zone organizations that would preferably bolster after-school programs than see an arrival to expanded quantities of children lingering around their stores every weekday evening.

- ***Are Quantifiable:*** Edge your objectives in a way that will let you know whether you have succeeded. Move from general dialect to particular dialect. Rather than saying, "Persuade more grown-ups to wind up coaches," evaluate the result you need: "Influence 25 working experts to end up noticeably center school tutors in the following a year." When our objectives are particular and quantifiable, we'll know whether we've succeeded.
- ***Engage, Change Conduct, or Force to Activity:*** Bringing issues to light might be a vital initial phase in your correspondence endeavors; however your objective at last looks to change conduct or constrain your groups of onlookers to make a particular move. Sustainability objectives regularly fall into the last classification (e.g., persuading State policymakers to support anti-bullying enactment, or influencing neighborhood organizations to finance grants for your late spring programs). Keep in mind: Communication objectives are not tactics!

Too regularly we wrongly think that a correspondence objective is to make a brochure or convey a quarterly pamphlet. Extraordinary thoughts, maybe, yet they're not objectives. These are strategies—thoughts that might possibly enable you to accomplish your objectives. You'll get to strategies in Step 5 of this arranging procedure, at the same time, for the time being, maintain your attention on what you need to achieve and what activities you might want others to take.

22.2.3. Stage 3: Identify Intended Audiences

Reason: To distinguish all potential target group fragments for every correspondence objective. To organize those gathering of people sections and learn however much as could be expected about them.

Since you've recognized your correspondence objectives, it's a great opportunity to distinguish the particular groups of onlookers you have to go after every objective. In your allows first years, you may have concentrated altogether on connecting with guardians, understudies, and instructors. Those gatherings of people may, in any case, be vital, however now you might need to concentrate on including group accomplices, policymakers, and region initiative to your correspondence endeavors. To put it another way, it's a great opportunity to make a rundown of who else should be at the table to help your sustainability objectives.

a. **Fragment Your Intended Audiences:** Fragmenting your groups of onlookers is a basic—and frequently disregarded—advance to making correspondence messages and materials that resound. For instance, how about we consider a run of the mill target group gather for sustainability: neighborhood organizations.

Suppose your correspondence objective is to acquire bolster for after-school programs. You know that neighborhood organizations are worried about truancy in the hours that take after the school day, and you can delineate the advantages of keeping youth possessed in the prior hours guardians return home from work. Your target group obviously would be neighborhood business pioneers, yet not all nearby business pioneers will have a similar level of enthusiasm for supporting your drive. To make the best correspondence procedure, fragment your more extensive target group into subgroups in view of their needs, values, as well as mission. For instance:

- Businesses that utilize secondary school understudies after school hours or over the summers (they show that they as of now esteem youth);
- Businesses in closeness to your school grounds (might be most specifically influenced by disposal of after-school programs);
- Business pioneers who have shown an enthusiasm for putting resources into your group's youngsters, especially around the issues of training and wellbeing;
- Business pioneers who are dynamic in your assembly of trade's group administrations board of trustees; or
- Business pioneers who have kids selected in your schools. Separating your general group of onlookers into fragments encourages you to create messages and materials custom-made to them. Next, Prioritize Your Audience Segments

You may now find that, rather than conveying to nearby organizations, you truly need to impart to five portions of your neighborhood business group. This may appear just as your work has expanded; however, it truly hasn't. You can organize your target group fragments and center your endeavors (deliberately!) on those gatherings at the highest priority on your rundown.

b. **At Long Last, Do A Little Homework:** Once you've distinguished and organized your target group portions, learn as much as you can about their missions, qualities, convictions, and needs. What wellsprings of data do they trust? It is safe to say that they are mindful of your drive's work, and would they say they are strong of it? How do your objectives for sustainability line up with their worries? What information may they have to see? What's the most ideal approach to contact them?

It's difficult to know excessively about your target groups, and the greater part of your discoveries will help you in making messages and materials that will probably induce them to make the coveted move. Research can be as straightforward as taking a gander at Web destinations for data on shared missions and needs. It can likewise be as straightforward as contacting accomplices to request data or to see whether there are built up connections that might be vital in your effort endeavors.

Since your objectives for sustainability may mean you have to extend your organizations, we urge you to allude to the Partnership Development area of this Celebration Kit. Those instruments can enable you to recognize solid prospects for businesses, open lines of correspondence, and build up those connections.

22.2.4. Stage 4: Develop and Pretest Messages

Reason: To create powerful messages that will reverberate with your target groups and constrain them to think, feel, or act in ways that help your correspondence objectives for sustainability.

The best messages bid to both a crowd of people's psyche and heart. This is on account of the mind looks for data that legitimately focuses a way for us as we decide; however, the heart must be moved for gatherings of people to really make a move. Powerful messages meet our groups of onlookers where they are by lining up with their qualities, convictions, needs, issues, and needs.

You and your group can create messages to be utilized as a part of numerous ways. Messages can pass on why your SS/HS activity genuinely matters. Your messages can fill in as the basic subjects featured and rehashed in your materials and exercises. They can be utilized as the premise of your arguments, introductions, one-on-one dialogs, or some other materials intended for your target groups. For instance, an all-encompassing message going through the majority of your materials may be that supporting your drive's sustainability endeavors is an interest in your group; the idea of venture could resound with a few diverse target groups, every one of whom are influenced by tight spending plans and constrained assets.

For target group individuals, viable messages should:

- Match their center esteems and convictions and show how your issue is in arrangement;
- Show the earnestness of your issue;
- Put a face on the issue or program;
- Provide verification of your esteem, either through quantitative information or narrative proof;
- Use dialect that bodes well and is free of language; and
- Include an invitation to take action that plainly advises what you might want them to do.

a. **Confining Your Message in Their Values:** Consider how your group of onlookers may answer this inquiry: "What sort of group would we like to live in?" Look at the table underneath for a little example of the sorts of qualities that may be most imperative to your gathering of people. As you consider what incentive to decide for your message outline, recollect that your answer ought to be unquestionably valid for your gathering of people. It ought to speak to the huge dream that they hold dear (Table 22.1).

Table 22.1: Types of Community

What kind of community do we want to live in? A community that is...

Source: Sprout (2017).

By encircling your message in their qualities, you set up a quick accord with groups of onlookers—even the individuals who may not put stock in aversion—expanding the odds that whatever is left of your message will be generally welcomed. Making esteem based edge for your message can fundamentally enhance the probability that your gathering of people will genuinely hear you since it shows that you and your group of onlookers share similar trusts in your group and its future.

b. **Express the Benefits:** Your gathering of people will need to know why your drive benefits them, and it can unmistakably characterize this in wording that will sound good to your group of onlookers. To anybody officially included with SS/HS, it might appear just as the best advantage originates from helping youth; however, that won't not be valid for potential accomplices. For instance:

- A neighborhood business may truly welcome the free advancement you can give them with signage at your occasions and specifies in your bulletins and on your Web webpage;
- Your adolescent equity framework may have more court cases than it can without much of a stretch handle and would have the capacity to relieve their burden by banding together with you to discover elective answers for adolescent offenses; or
- Mental wellbeing specialist organizations may profit by the expanded customer base that could originate from cooperating with you. Defeat Their Barriers Your gathering of people will most likely have reservations about what you are putting forth, and these boundaries should be overcome before your group of onlookers will state yes. A crowd of people may feel that you are asking something that appears to be excessively costly or too tedious. It may not appear to be reasonable or significant to them. This is your chance to clarify how you are approaching something that is simple for them to do.

As you are building up the message parts that both layout benefits and defeat obstructions, bear in mind to survey the discoveries you caught in Step 3—it will enable you to maintain your attention on your gathering of people's needs, needs, and inclinations as you create dialect. Both of these message segments are additionally a characteristic fit for including particular quantitative information, narrative stories, or social math that will reverberate with your group of onlookers and help present your defense.

For instance, your school board may have made an open sense of duty regarding trimming the financial plan while expanding test scores. They might be hesitant to add aversion administrations to the financial plan, and that is a boundary. In this way, you may need your messages to board individuals to pass on that avoidance administrations can accommodate more tranquil classroom atmospheres in which all understudies have an enhanced chance to learn and develop. You may likewise utilize information to exhibit that your projects are keeping more children in school—and in many States that implies all the more financing dollars consistently, for each school.

c. **Invitation to Take Action:** Since you have conveyed a convincing and modified message, what might you like your crowd to do next? Keep in mind that an extraordinary invitation to take action does not need to move mountains at the same time. On the off chance that your group of onlookers has never known about your program, a suggestion to take action that demands a gift of products or administrations may not be as practical as a demand for a follow-up meeting to broadly expound on your answer.

Here is a rundown of normal invitations to take action that may help control your reasoning:

- Take your next telephone call;
- Write a letter/get the telephone for your sake;
- Introduce you to another person;
- Invite you to talk before another gathering;
- Serve on a warning board;
- Come to a gathering;
- Provide administrations to customers;
- Provide financing for sustainability; or
- Change a conduct. In creating messages, recollect: It's not what you need to let them know, it's what they can hear. This implies adjusting your message to their qualities and convictions, beating hindrances, and delineating benefits in ways that sound good to them. Pretesting Messages Before pushing ahead to create materials or exercises, we ask you to pretest your messages with only a couple of individuals from your target groups at whatever point conceivable to decide whether the messages are powerful. Request their information, and, in the event that you find that your message doesn't have the effect you sought after, utilize this criticism to change your message before you continue to creating materials or introductions. You may have just a single opportunity to establish a decent connection with your gathering of people, and pretesting your message will enable you to take care of business.

22.2.5. Stage 5: Select Channels, Activities, Materials, and Partnerships

Reason: To recognize the most proper roads for conveying your messages to your target groups.

Deciding the best way(s) to contact your target groups implies backpedaling to the work you did in Step 3. Invigorate your memory:

- What wellsprings of data do they trust?
- Who or what may propel them to make the coveted move for your sake?
- How would they want to get your data?
- How and where do your target groups invest energy? Whatever city affiliations, confidence based gatherings, or different businesses do they have a place? Answers to these sorts of inquiries will enable you to convey your messages viably; however, there are different contemplations too. You have to decide how you can guarantee your correspondence will be:
- Considered to be socially proper;
- Delivered to your target groups in an opportune way;
- Aligned with your financial plan and assets; and
- Tailored to your target groups' general observations about youth savagery avoidance and bolster administrations. This progression guides you to choose the channels, exercises, and materials you will use to convey your message. The following are only a couple of illustrations that might be applicable to your correspondence get ready for sustainability and can assist animate exchange with your accomplices. Channels Communication channels convey your messages to the target groups. Channels take many structures, and there is a nearly boundless rundown of potential outcomes. A few cases of channels include:
- Individuals (you, your champions)
- Radio stations
- Web sites
- Email
- Malls
- Schools Recreation focuses
- Supermarkets
- TV slots
- Newspapers

- Community focuses
- Laundromats Parks
- Libraries
- Nonprofit businesses
- Restaurants

a. **Exercises, Events, and Materials:** Exercises and materials are the strategies and devices used to send messages through the channels.

A few illustrations include:

- Exercises and Events Open houses
- News gatherings
- Community occasions
- Conferences
- One-on-one gatherings
- Public testimony
- Award services
- Presentations to bunches
- Family days

b. **Materials**

- Factsheets
- PowerPoint introductions
- Webs
- Open administration declarations/recordings Reports
- Community report card
- Flyers and leaflets
- Posters
- Bookmarks
- Other marked special things

c. **How Often, and in How Many Ways, Should You Deliver Your Message?:** Envision a youthful kid approaching her parent for another toy. That young lady could never approach only once for the toy—she would ask over and again, and she would get progressively imaginative with how she would ask her folks. She may ask her mom toward the beginning of the day, draw a photo of the new toy and put it on the icebox toward the evening, and complete the day by getting her different kind to ask their father while he is unwinding in the wake of a prolonged day. Guardians may flinch at the possibility of this; however, you need to commend the perseverance and inventiveness of these youthful advertisers who are as a rule effective. The same principle applies to you and your initiative. For example, your school district is far more likely to support your sustainability efforts if they have received regular updates from you and your staff for the past 3 years, rather than one PowerPoint presentation when you first won the grant. Similarly, a simple PowerPoint presentation might not convey the full impact of your SS/HS initiative to the board; a student or school resource officer telling the story of how the initiative has affected his or her life positively could be far more powerful.

This principle also applies to the business community, particularly in this challenging economic time. Local business leaders are more likely to assist your sustainability efforts if you have regularly communicated the benefits their business receives because of the SS/HS initiative. One-on-one meetings, newsletters, or perhaps an invitation to an event could communicate how their business benefits from lower truancy rates and a better-educated and well-behaved workforce. Increased graduation and college attendance rates could also translate into better-paying jobs for these SS/HS students, which mean the students will have more money to spend in local stores down the road.

As you develop your messages and materials, we encourage you to incorporate meaningful evaluation data that help tell your story in a way that will resonate with your intended audiences. For a more in-depth examination of this critical communication tool, please refer to the Communicating Evaluation Data section in this Celebration Kit.

22.2.6. Stage 6: Develop an Action Plan

Reason: To figure out where, when, and how each assignment will be done to effectively actualize the correspondence get ready for sustainability. To dole out duty regarding each piece of your arrangement.

The best correspondence design on the planet is only an intriguing archive without an activity design. This progression ensures that the majority of the diligent work you've done sees the light of day.

An incredible activity design is great task administration and makes the best utilization of everybody's opportunity. We support shared investment in putting your arrangement on its feet. Duty from staff individuals, accomplices, volunteers, and different champions tells your group that there is wide help for your objectives—and it may urge potential new accomplices to help your prosperity too.

Your activity design can be as basic or as mind-boggling as your needs require, at the same time, at least, your activity design ought to have the accompanying fundamental data:

Major exercises and errands;

- Intended date for finishing each errand; and
- Name of the individual in charge of guaranteeing that each undertaking is finished. Table 22.2 is an example of one assignment in an essential activity design.

Table 22.2: Communication Planning

Task/Subtask

Source: Sprout (2017).

22.2.7. Stage 7: Develop and Pretest Materials

Reason: To guarantee that your materials will reverberate with target groups before they are delivered and circulated.

Correspondence and social promoting experts have learned numerous lessons throughout the years about what influences a correspondence to program or social advertising effort viable. A standout amongst the most imperative lessons is the benefit of pretesting messages and materials. Having agent individuals from the target group audit your items previously they are created or appropriated is fundamental to the accomplishment of your correspondence.

In case you're enticed to skip pretesting, consider the potential traps. Keep in mind that, we may think we completely comprehend our target group; however, we presumably don't. Pretesting can spare us from committing expensive errors.

Pretesting encourages you:

- Identify whether your messages and materials have any real defects;
- Explore elective messages and materials;
- Fine-tune your messages and materials. Regularly, creating correspondence materials is a four-advance process:

1. Develop model;
2. Review and pretest;
3. Revise and refine; and
4. Create.

There are a few approaches to pretest messages and materials, including:

- Surveys;
- Focus gatherings;
- One-on-one meetings; and
- Advisory sheets. As a matter of fact, in case you're arranging a vis-à-vis meeting with a person who speaks to a potential accomplice for sustainability, it's entirely hard to pretest your message and materials with an individual from your target group. In such a case, it's a smart thought to practice your introduction and demonstrate your pieces of literature to somebody who isn't associated with your drive's work. While you will most likely be unable to check the effect of your message and materials, you'll in any event have the capacity to ensure that you're introducing your data in a way that is simple for anybody to get it. Keep in

mind—while making correspondence materials, influence sure to incorporate time for pretesting in your activity to design.

22.2.8. Stage 8: Implement, Evaluate, and Modify Plan

Reason: To decide whether the arrangement is meeting your correspondence objective and, if not, to make suitable and opportune alterations to guarantee achievement.

Dealing with a correspondence design or social showcasing effort is a continuous procedure. Execution is just the start. Consider your correspondence design a "living record" that is reexamined and tweaked after some time.

To guarantee that your correspondence and social showcasing endeavors are powerful and fruitful, you should determine how and when you will assess and, if fundamental, adjust your technique and activity design. As you reveal your correspondence design, we urge you to make an input instrument to screen your arrangement's prosperity. In the event that something in your program or battle isn't working, don't simply continue through to the end: Go ahead and make the adjustments important to get back on track.

Recollect your objectives (Step 2) as you evaluate the achievement of your endeavors. Since your objectives were quantifiable, you will know whether your arrangement is working. For example, if your objective was to induce neighborhood organizations to give subsidizing to grants for your late spring programs, you will know you've succeeded on the off chance that you get the financing.

Be that as it may, imagine a scenario in which you're not seeing the expected aftereffects of your arrangement. Utilize this progression to:

- Determine which messages are not reverberating with your target groups;
- Identify the channels, materials, exercises, and businesses that are not pushing you to reach and draw in your target groups; and
- Identify the snags that you had not expected. Your assessment may likewise direct you to new gatherings of people that can enable you to achieve your objectives—groups of onlookers you hadn't perceived when you began this arranging procedure. At the point when this happens, you will need to begin again with Step 3, learning as much as you can about these new gatherings of people, at that point making and pretesting messages and materials for them. Congrats! By working through these means,

you and your accomplices have made an authentic sense of duty regarding the achievement and long haul sustainability of your SS/HS activity. You are presently figuring deliberately to:

- Make the vast majority of your accessible assets and your substances;
- Set reasonable and noteworthy correspondence objectives;
- Identify the particular target group sections that are well on the way to help your endeavors;
- Craft messages that resound with those gatherings of people;
- Deliver those messages in ways that address the issues and inclinations of your gatherings of people. Moreover, you perceive that the best designs are justified regardless of close to the paper on which they're composed unless they are actualized. Presently, you are prepared to:
- Make an activity arrange for that doles out errands and due dates;
- Create, pretest, and deliver any materials;
- Implement, assess, and change your arrangement.

22.3. CASE EXAMPLE FOR SOCIAL MEDIA FOR A NON-PROFIT ORGANIZATION

22.3.1. Introduction

Philanthropic businesses have turned out to be progressively social over the previous decade, regardless of whether they are developing a devoted Facebook following or pioneering a trail on new systems, for example, SnapChat. While part of online networking's underlying interest was the possibility to contact your crowd without taking advantage of a constrained spending plan, the developing prevalence and evolving at the highest point of supporters' newsfeeds. All things considered, philanthropic businesses have never been a gathering to down from a test.

Web-based social networking is a chance to recount your story, draw in with supporters and get comes about, which implies there's no time like the present to make or revive your online networking methodology. We requested that five specialists say something regarding how to build up a not-for-profit web-based social networking procedure and demonstrate the estimation of online networking to your association. As we layout the means to:

- 01 – Define Social's Role in Your Communications Strategy;

- 02 – Determine What You Want to Accomplish;
- 03 – Identify Your Target Audience;
- 04 – Choose the Right Networks;
- 05 – Create a Content Strategy;
- 06 – Put Engagement First;
- 07 – Empower Your Advocates & Cross-Promote Your Content;
- 08 – Track & Measure Your Results.

a. **Define Social's Role in Your Communications Strategy:** "Web-based social networking shouldn't be out there alone; it ought to be coordinated and lined up with your vital objectives and target gatherings of people," said Beth Kanter, Nonprofit Consultant and Author of Beth's Blog, one of the longest running and most famous online journals about philanthropics and web-based social networking. "The part that it can play extremely well is engagement, particularly benefactor engagement and contributor maintenance, motivating individuals to focus—catching their consideration."

The moves you will make and the substance you will make on social should attach back to your general correspondences and computerized system to your site and blog, email showcasing, occasion advancements and some other substance you share on the web.

b. **Determine What You Want to Accomplish:** As a major aspect of a not-for-profit group, you're no odder to having a mission. Regardless of whether to dispose of destitution or secure a jeopardized species, you comprehend what your general charge is. Likewise, your social objectives ought to be clear (Figure 22.1). To give you a feeling of normal correspondences objectives that can illuminate your social endeavors, in 2017, not-for-profit communicators distinguished these four objectives as needs for the year:

- Community engagement and training
- Brand building and notoriety administration
- Program enrollment

Figure 22.1: Social media twitter post.

Source: Sprout (2017).

Web-based social networking gives chances to association that different channels may not. Chara Odhner, Senior Copywriter and part of the social group at philanthropy: water, echoes the significance of utilizing social for group engagement.

"As an association that is had a $0 promotion and showcasing spending plan since the very first moment, we put resources into social from the get-go," Odhner said. "Also, we adore social for what it specializes in: fortifying our group of supporters."

Once you've decided your objectives, choose how you will gauge achievement and what measurements you will track to evaluate the result of an objective, you may gauge accomplishment by reactions to your posts and engagement in social discourses, volume of posts utilizing a particular hashtag and interest in online battles driven by social.

c. **Identify Your Target Audience:** "A great deal of charities say, 'well, our group of onlookers is the overall population,' yet in the event that you believe you're addressing the overall population, you're most likely addressing no one," said Bridgett Colling, Director of Content Marketing at See3 Communications, a computerized office for philanthropies and social causes.

Colling suggests that not-for-profits create group of onlooker's personas, which are portrayals of your optimal supporters in view of a mix of statistic information and data about individual individuals from your intended interest group.

To find out about the statistic cosmetics of your present online networking tailing, you can utilize local investigation on stages like Twitter and Facebook or utilize a web-based social networking administration instrument. When you begin a free trial of Sprout Social and interface your profiles, your groups of onlookers' statistic information will start to populate, which you can see effectively in your dashboard or fare as a report.

When you have a thought of who's as of now tailing you on the web, about what makes a difference to them when supporting your cause.

You can make different personas that speak to supporters in board and junior board, your benefactors and then some. Simply ahead and give every persona a name and also an exhaustive statistic foundation, at that point include particular insights about what this individual thinks about, what their normal day resembles, who they trust and the sky is the limit from there. Nancy Schwartz's blog has an accommodating well-ordered manual for making personas (Figure 22.2).

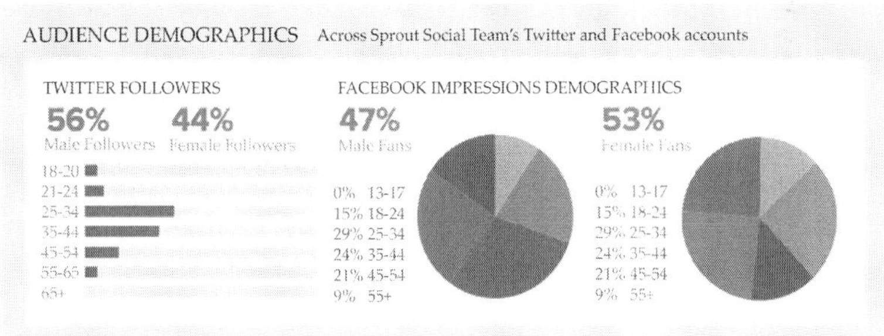

Figure 22.2: Audience demographic.

Source: Sprout (2017).

d. **Choose the Right Networks:** When you have recognized your intended interest group, ponder web-based social networking socioeconomics to discover where they are dynamic on the web, and center your social—and who doesn't?—don't spread yourself too thin by agonizing over receiving each system that surfaces. Complete a magnificent activity on a few systems with a huge potential for contacting your gathering of people rather than a fair occupation on five.

To give you a thought of which systems are most well known among charities, Facebook remains No. 1, with 96% of philanthropic communicators positioning it in their best three online networking locales; Twitter and YouTube take spots two and three, separately, as per the 2015 Nonprofit Communications Trends Report.

e. **Create a Content Strategy:** While making a substance technique, concentrate on narrating that communicates as the need should arise.

1. **What to Share:** Begin by inspecting past presents on observe what has performed well and what kinds of substance get your adherents locked in. Try not to be reluctant to solicit supporters what compose from stories they might want to see from your association. You can accumulate this data face to face, by means of an email review or by asking via web-based networking media. Create content classifications for online networking that you can go through all the time, for example, volunteer spotlights, customer stories, and how-to-dos.

Kivi Leroux-Miller, Nonprofit Consultant and President of Nonprofit Marketing Guide, proposes two inquiries to control your substance creation: "What issues do individuals have in their own lives when endeavoring to experience the qualities they share with your association? What tips or instruments would you be able to give them that make their lives less demanding as they endeavor to be a superior naturalist, creature sweetheart, parent, and so on?" (Figure 22.3).

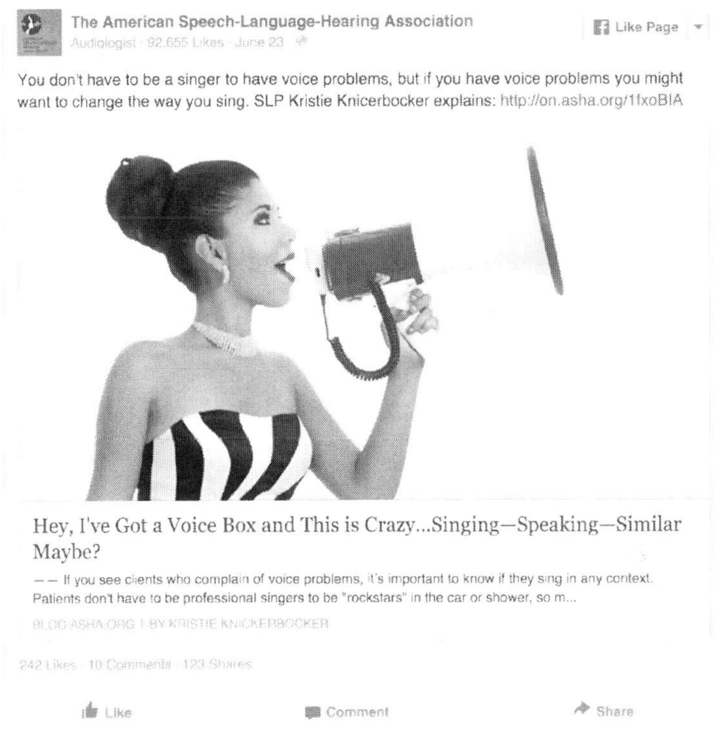

Figure 22.3: Social media Facebook post.

Source: Sprout (2017).

2. **Sorts of Content:** From pictures and illustrations to short recordings and content, your association should share an assortment of posts over its systems. Visual substance can frequently achieve enthusiastic triggers in a way that words alone can't. There are a lot of free instruments for picture creation that can enable you to make delightful illustrations, yet don't stress over making each picture exceedingly cleaned. Catching depictions and short recordings on your telephone empowers you to share capable minutes from occasions, and those kinds of posts can give your supporters a significant look in the background of your association.

While each post needn't bother with a connection, online networking can be a useful device for guiding individuals to your site, blog or mailing list. Gathering pledges Coach and Author Marc Pitman says that one regular error is keeping your whole online networking nearness via web-based networking media.

"Facebook could change the tenets; Twitter could close down," Pitman said. "You ought to have a solid blend of substance that focuses back to your site and helps manufacture your email list where you give away supportive substance—a moral fix. At that point, you can track what number of individuals are originating from social and joining or making a move."

3. **Posting-Frequency:** Making a web-based social networking content timetable will enable you to plan and calendar your online networking posts ahead of time. On the off chance that you are working with a group, or connecting with others in your association for photographs or data, a substance logbook will enable you to remain over arranging efforts and getting ready materials ahead of time. Decide a posting recurrence and rhythm that will keep your association before your group of onlookers while abandoning you an opportunity to deal with your group, share newsworthy substance and measure your effect.

f. **Put Engagement First:** Social media is far more than a publishing platform—it's a place to capture people's attention, connect with supporters and build communities.

"Social media is designed for two-way conversation, and that's when it is most powerful."

—Chara Odhner, Charity: Water

"The most powerful thing about social media is something many companies and organizations often forget: It's social," Odhner said. "Instead, many brands use social media as a broadcasting platform."

With limited time to spend on social, engagement should be a priority. Respond to questions, comments, and posts tagging your organization, and look for relevant hashtags to find new conversations to join. While your tone may be more formal on other outlets, social media is a particularly good place to cultivate a personable brand voice that helps supporters feel connected. Don't be afraid to use humor to connect either.

"Part of what we're trying to do with engagement is show donors that we see them as more than an ATM," Pitman said. "The social media accounts that show humanity seem to get noted more, particularly when you give quick or humorous responses" (Figure 22.4).

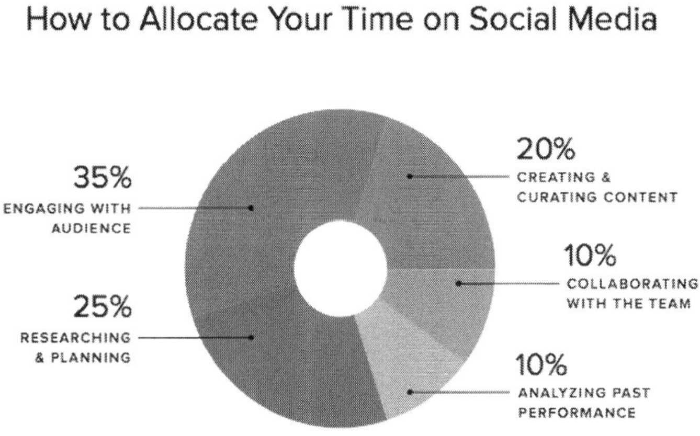

Figure 22.4: Time allocation of social media.

Source: Sprout (2017).

g. **Empower Your Advocates and Cross-Promote Your Content:** Indeed, even the most eager advocates of your association may not understand that social sharing is an intense way that they can bring issues to light for your motivation. On the off chance that you are making or increase your essence on a particular system, make it known to the greater part of your constituents that they can associate with you there. Demonstrate to them the benefit of following—the stories, tips, and pictures they can hope to see—and teach them about the most ideal approaches to indicate bolster. Try not to be hesitant to cross advance your web-based social networking content on different channels. Specialists gave a couple of proposals to get your supporters talking.

1. **Begin a Private Group:** "Welcome them to join a private gathering or another instrument of correspondence so you can 'shake 'n' move,'" Kanter said. "Everybody likes to be on a triumphant group and to give a shout out to the group, so you require that sort of association. Make it super simple: Provide test posts for Facebook, test Tweets, and so forth."

2. **Advance by Means of Other Communication Channels:** "Bolster your social with email, telephone and different methods for correspondence," Pitman said. "Send a connection to board individuals, and request that they like it share it, remark and reveal to you what they think. Be that as it may, you need

them to cooperate with you, let them know. Individuals can't read our psyches. It's not evident to them; their universe doesn't rotate around our philanthropic or our online networking effort."

3. **Give Sample Posts:** "Make it simple to share by sending an email with pre-written Tweets and illustrations individuals can share when you're propelling a major crusade, revealing to them one of the ways they can bolster you is to share on Facebook, Twitter or whatever system you believe they're most dynamic on," Colling said. "Make it simple for individuals. Give them alternatives. What's more, reveal to them that sharing is significant."

h. **Track and Measure Your Results:** Advancing toward your objectives. Exhibiting results will demonstrate your initiative the significance of putting time and assets into web-based social networking, and it will enable you to modify your substance procedure to concentrate on the sort of substance that performs best.

"If it doesn't get results, don't do it. You don't get impact by wishful thinking."—Beth Kanter

Kanter suggests that you look not only at what content is getting a good rate of return but also at what takes less time to get that return.

Google Analytics to see how many social referrals are getting people to your website.

"Even better, if you put goal tracking in your Google Analytics, you can see how many referrals actually led to someone completing a donation," she said.

On the off chance that setting aside a few minutes to run reports is a test; this is only one of numerous regions where a web-based social networking administration instrument will help. You can utilize Sprout to run week by week or month to month introduction prepared reports on everything from sent message execution to group of onlooker's changes to engagement. You can likewise utilize data from interpersonal organizations' local investigation (Twitter Analytics, Facebook Insights, and YouTube Analytics) to make your own particular reports. For a reasonable breakdown of how to utilize local examination, look at our post on the online networking measurements that issue.

22.3.2. Conclusion

Most importantly, make sure to tune in and make a propensity for technique and estimation. After some time, you will realize what content gets your group talking and how to calibrate your charitable's online networking system to get the most ideal outcomes. We'll abandon you with this one last suggestion:

"It's important to set reasonable expectations and know that social media and content marketing is a long game. You're not going to put one post out there and instantly get hundreds of dollars in donations. Like any good relationship, it's all about communication over time."—Bridgett Colling.

CHAPTER 23

Reporting, Global Communications, and Certifications

23.1. REPORTING AND COMMUNICATIONS

Sustainability announcing is a fundamental initial phase in the process by which businesses make a move to make a really feasible worldwide economy; this conviction has dependably supported GRI's main goal. Be that as it may, announcing is not static – it advances, through improvements like rising worldwide difficulties, new innovations and changing partner interests – and organizations need to think about this.

The Corporate Leadership Group (CLG) on Reporting 2025 was met by GRI to advance the worldwide dialog on the fate of sustainability revealing and divulgence, and to analyze how the act of announcing is developing.

Thirteen forward-looking organizations met up to investigate the patterns they predict will be more important later on. From the 17 patterns GRI displayed at the Sustainability and Reporting Trends Forum held in November 2015, the four were researched and better comprehends the impact they would have all alone organizations and sustainability reports (Figure 23.1).

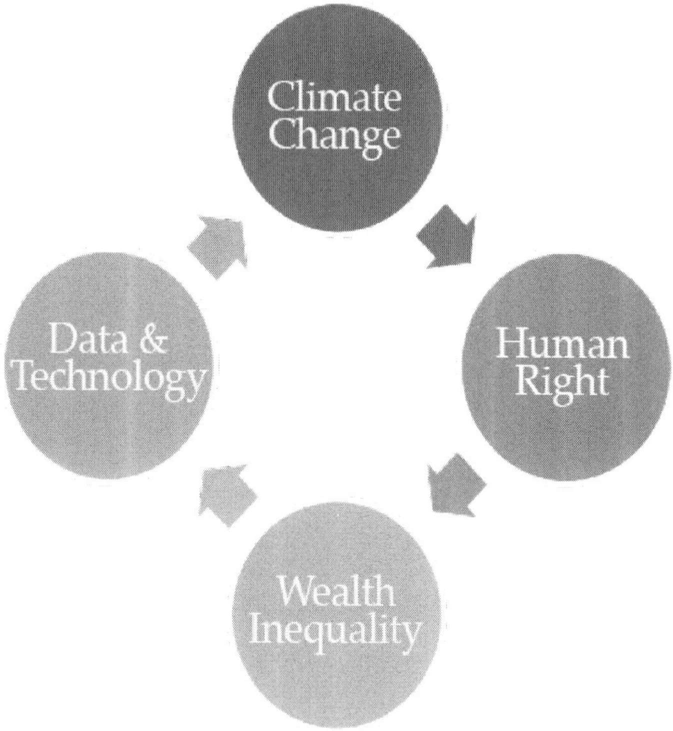

Figure 23.1: Reporting and communication areas.

Source: GRI (2017).

23.2. EXPLORING FUTURE TRENDS

To increase new bits of knowledge and a more profound comprehension of the potential impacts of these patterns all alone organizations and sustainability reports, the four patterns were dissected in four angles.

Diverse specialists were welcome to impart to the gathering their considerations on the fate of announcing, and Sustainability gave content and encouraged the talks. In the Labs, the gathering adopted an organized strategy to investigating the patterns, covering the accompanying four viewpoints (Figure 23.2).

Figure 23.2: Reporting process.

Source: GRI (2017).

CHAPTER 24

Climate Change

24.1. CURRENT LANDSCAPE

Environmental change has been on the spotlight on the world agenda as well as embedded in the most advanced implementation systems for sustainability and corporate communications. In 2016, the World Economic Forum named the inability to alleviate and adjust to environmental change the top worldwide hazard regarding sway. Atmosphere highlights noticeably in the 17 Sustainable Development Goals (SDGs), which are now accumulating noteworthy intrigue and early take-up. Many revealing businesses as of now unveil data on environmental change; the Organization for Economic Co-operation and Development (OECD) and the Climate Disclosure Standards Board (CDSB) took supply of environmental change divulgence in the G20 nations to locate that corporate announcing practices are enhancing quickly, with practice that was considered spearheading in 2010 getting to be noticeably standard in 2015.

The key points made at COP21 in Paris in December 2015 brought about:

- A more tightly atmosphere strategy condition.
- A long-haul objective being set for worldwide-carbonization.
- Momentum in financial specialist enthusiasm for the issue, so far playing out by segment lines.
- The sureness that administrations can't accomplish worldwide targets alone.

The following stride of taking load of governments' Intended Nationally Determined Contributions (INDCs) to environmental change is in progress.

The various related activities additionally raised in 2016 are:

- The OECD has been appointed to evaluate the past responsibility of a $100 billion store by 2020 to help creating nations adapt to environmental change.
- The Breakthrough Energy Coalition of innovation pioneers is giving $1 billion to development to people in general research pipeline.
- 120+ organizations have focused on creating science-based targets.

24.2. DRIVERS FOR CORPORATE PERFORMANCE ON CLIMATE CHANGE

Corporate authority on environmental change should be driven in various regions. The top variables driving organizations to address environmental change are:

- Using elective and sustainable power source.
- New advancements/development.
- Setting targets and having an atmosphere methodology.
- Communicating and supporting on atmosphere.
- Supportive administration.
- Leadership from the top.
- Committing and conveying genuine outcomes.
- Reviewing items and administrations for atmosphere impacts.

"The Bank's commitment to the Paris agreement is an essential piece of our need setting. Atmosphere related back, green fund, green securities are issues where the Bank has been centering for a considerable length of time, and positively that concentration is not going to leave." – Parminder Plahe, European Investment Bank (EIB).

Besides, environmental change is being regarded money related: The Financial Stability Board presented the Task Force on Climate-related Financial Disclosures to create intentional predictable atmosphere related monetary hazard exposures for organizations to give the correct data to speculators, moneylenders, back up plans and different partners.

24.3. IMPLICATIONS FOR BUSINESS

There was a clear agreement in the gathering that it is not a matter of if business ought to or can follow up on environmental change however how and how quick they convey change. The accompanying focuses were exhibited in the Lab for the CLG to talk about:

- Companies are arrangement suppliers: they are relied upon to be a piece of the arrangements, from new energy models to efficiencies in the generation and dispersion of products.
- Climate change is interrelated with different issues: how does environmental change communicate with and affect different issues, for example, human rights?
- Clear, yearning, science-based targets are required: new pointers on environmental change impacts and related externalities are developing yet deficient even with concurred worldwide cutoff points.
- Linking to INDCs is vital: more prominent organization and nation engagement are relied upon to be required with the Paris understanding.

"Environmental change impacts horticulture. For our sourcing, the test is the manner by which to draw in ranchers and those included in rural creation on environmental change. Our report ought to be more unequivocal, more realistic and clearer in its objectives to make it more conceivable for our own particular partners, particularly for our accomplices in horticulture." – Sol Beatriz Arango, Grupo Nutresa.

24.4. EXAMPLES: BUSINESS ADDRESSING CLIMATE CHANGE

Cases of how organizations incorporate atmosphere related data in their sustainability reports:

- CEMEX has detailed widely on environmental change, ordering advancement and future course into an independent position, which incorporates advance against targets and flags a move to inner carbon valuing.
- Power organization CLP set an objective in 2007 – which still stands – to lessen the force of carbon outflows by around 75% by 2050.

- Food and refreshment organization PepsiCo is working towards a flat out lessening in ozone-harming substance emanations over the business and furthermore utilizes the focal point of the PepsiCo esteem bind to demonstrate the limits of this material issue.
- EIB, the world's biggest multilateral lender, has focused on straightforward and nitty-gritty giving an account of the atmosphere advantages of its speculations, by catching the GHG emanations of 95% of its venture ventures, while screening out activities whose GHG outflows surpass a stringent edge level.
- Renewable energy bunch Gamesa plans to accomplish carbon lack of bias by 2025. This test includes singular duties on the estimation, lessening and additionally counterbalancing of the CO_2 starting from the Group's immediate and roundabout discharges from 2015 to2025.

To see other driving cases, see the GlobeScan/Sustainability 2015 Climate Survey, which recognizes driving corporate environmental change objectives. One territory where there is wide accord is setting aggressive targets: RE100 is turned out to be an intense activity, through which a developing rundown of organizations have focused on 100% sustainable power.

24.5. PRIORITIES AND NEXT STEPS

Priorities for exploring and improving their reporting on climate change:

- Contextualize and disguise the test of environmental change more: Learn what you can from best practice yet guarantee you think about your materiality examination and the parts of environmental change most important to your division and association and enhance information, measurements and announcing as needs, e.g.:
  - If the accentuation is on your inventory network, your need may comprehend the dangers in your store network and computing your extension three emanations.
  - If your items and administrations impact the energy utilize and discharges of others, their plan and conveyance might be generally intense.
- Aim to comprehend the interconnections, or nexus, of environmental change and different issues.

- Set and accomplish long haul (past 2020), science-based targets (endeavour to get at 'what's coming to you'), which likewise separates organizations.
- Determine how to adjust the interests and exercises of revealing businesses on governments, given national anticipate environmental change and the multinational way of many announcing businesses.
- Develop a reasonable and compact story of the helplessness of your association in light of environmental change and the common test it presents.
- Focus on conduct change: what will people and businesses need to do any other way to address the difficulties around energy utilization and emanations? "In a resource-constrained world, CEMEX understands the importance of reducing emissions and the need to quickly incorporate clean energies in the elaboration of our product. We understood we had to come to solutions quickly to contribute to a cleaner product and decrease our footprint." – Martha Herrera, CSR Director, CEMEX.

CHAPTER 25

Human Rights

25.1. CURRENT LANDSCAPE

Human rights insurance is a key and developing societal concern, especially in light of on-going clashes (e.g., in Syria) and relocation and displaced person streams. For the greater part of the advance business has made human rights manhandle still happen. Many organizations are as yet pondering worries in their supply chains, specifically – be it the unprotected transient workers in the palm oil or fish supply chains in Asia, the worry for Syrian outcast youngsters working in clothing production lines in Turkey, or present-day bondage in the UK. We see significant improvements with respect to human rights moving towards national execution and responsibility:

- Human rights are implanted over the Sustainable Development Goals (SDGs): the Danish Institute for Human Rights proposes we can "… utilize human rights as a driver for understanding the SDGs… and utilize the SDGs to acknowledge human rights."
- Various countries – presently including the UK, Italy, Spain, and Colombia – have created National Action Plans particularly for human rights.
- The UK's Modern Slavery Act was viewed as historic point enactment, influencing an expected 17,000 organizations: huge UK-based firms are required to distribute confirm on their sites showing that their operations are not dependent on universal subjugation.

25.2. DRIVERS OF CORPORATE PERFORMANCE ON HUMAN RIGHTS

Phil Bloomer, Director of the Business and Human Rights Resource Center, specified two patterns that will be basic to business operations and show the interconnectivity of the patterns:

- The ascent of hyper-disparity and chance of shared thriving: Business assumes a crucial part in the route in which riches is made, additionally in how riches is conveyed. Regarding and securing human rights is major to the equivalent conveyance of assets.
- The compounding biological emergency: The need to move to a low carbon economy makes a chance to discover arrangements that additionally advantage the most powerless – those frequently left unprotected earth or naturally likewise regularly encounter an absence of human rights and social insurance, particularly in long and complex supply chains.

25.3. IMPLICATIONS FOR BUSINESS

Desires of corporate providing details regarding the numerous features of human rights are developing and the issue may quickly develop:

- Human rights due persistence is presently the normal least.
- Benchmarking and positioning will empower simple and fast examinations between businesses.
- Investors, rating offices and controllers are progressively looking for this data.
- Greater required straightforwardness and due steadiness: This is obvious all around with requests from common society and expanding arrangement and control.
- An emphasis on work rights as basic: Modern subjection is another type of extreme work manhandles, and is prompting a more extensive development from an attention on review and consistence to due tirelessness and cooperation.
- Stakeholders are stressing the requirement for least living wages, hence influencing brands and unions, modern aggregate understandings and possibly making compensation pre-aggressive. Wellbeing and security, flexibility of affiliation and

the privilege to aggregate haggling likewise remain concerns.

- Conflict over common asset riches: Land rights are progressively debated and, as water shortage declines, question is required to develop. Extraordinary air contamination is likewise picking up consideration. As the world moves to low carbon economies, there is support to guarantee that the move is quick and reasonable. At long last, corporatc duty and assessment asylums have a part to play to guarantee expense is paid where asset riches are separated.
- Emerging zones of solid corporate promotion: various companies have talked and acted to end segregation by sex or sexual introduction, for example, while in a few divisions –, for example, assembling, extractives, and innovation – support for the regard of fundamental rights all through the inventory network remains a need.

"Ensuring that compensation and remuneration programs do respect and improve minimum wages established in different markets and countries, or reducing the gender wage gap becomes a key challenge for multi-located companies like ours." – Isaac Ruiz, CSR Director, Gamesa.

25.4. EXAMPLES: BUSINESS ADDRESSING HUMAN RIGHTS

The cases of how organizations incorporating human rights-related data in their sustainability reports are:

- ICT organization Telefonica concentrates on the human rights issues around computerized trust, which incorporate security and security, flexibility of expression and youngster insurance. Telefonica is chipping away at the execution of its 10 Guiding Principles on Privacy and Freedom of Expression.
- Grupo Nutresa has a structure of activity – a reasonable independent correspondence on where the organization remains on human rights and how they conform to their human rights arrangement.
- Renewable energy organization Gamesa and IndustriALL Global Union consented to a worldwide structure arrangement (GFA) including a pledge to regard fundamental social rights, human rights and common freedoms, level with circumstances, justify

and non-separation.

- Indian iron and steel maker JSW trusts a various workforce is a key element for a creative biological community. They exhibit the breakdown of the workforce by age, sex, classification, level, and turnover.

25.5. PRIORITIES AND NEXT STEPS

- Better comprehend human rights, not just as a material issue for some organizations, however a greater comprehension of striking nature and which rights are in danger of the most serious effects through the organization's exercises and business connections.
- Consider doing this through human rights affect evaluations.
- Consider receiving the UN Guiding Principles on Business and Human Rights Reporting Framework.
- Building mindfulness and spreading learning all through an organization and its inventory network is crucial.
- Companies are engaged with investigating how their HR offices can help serve human rights.
- Integrating regard for human rights all through the production network is viewed as the greatest test.
- Companies are additionally intrigued by stages to share store network data, as a huge number of providers in any one inventory network can't be checked or inspected by a solitary association viably.
- Multi-partner coordinated effort for aggregate activity around a range of human rights, regardless of whether ware particular or a neighborhood area, is testing yet worth handling.
- Supply chain chance investigation work and inspecting is an initial step to distinguishing dangers in human rights.

CHAPTER 26

Wealth Inequality

26.1. CURRENT LANDSCAPE

As per Oxfam, the wealthiest 1% on the planet now has more riches than whatever is left of the world consolidated. In the sustenance and drink industry, for instance, 60% of ranch workers live in neediness. Little-scale ranchers create 80% of the world's nourishment, yet together with horticultural laborers, they make up the lion's share of the individuals who are ravenous.

The OECD characterizes riches imbalance as "monetary disparity… the distinction in how resources, riches, or pay are circulated among people as well as populations." It is likewise portrayed as the hole amongst rich and poor, pay imbalance, riches divergence, riches and pay contrasts, and the riches hole.

Tending to riches disparity is progressively observed as a crucial essential of a feasible economy, with various key partners and specialists concentrating on the issue and how it interfaces with get to monetary assets, instruction, wellbeing, and so forth. With the ascent of hyper imbalance, and every one of the dangers it presents, there is a significant chance to concur and re-adjust on what is more equivalent and shared success.

Examining the sustainability announcing of a few organizations, it is clear that riches imbalance is infrequently tended to straightforwardly as a pattern or some portion of the more extensive working setting impacting business. There are signs that more organizations are perceiving the potential – and desire – for them to impact financial difficulties through particular business exercises, for example, work, wage levels, and proportions and neighborhood sourcing.

Drivers of corporate performance on wealth inequality as indicated by Ioan Nemes, Policy Lead CSR, and Sustainable Agriculture at Oxfam Novib, there are two key measurements of driving worry for riches imbalance and picking up organizations' consideration:

- An expanding offer of the estimation of our worldwide economy is going to capital, as opposed to work, which addresses the difference between lower wage specialists and high riches people. Many expense frameworks don't guarantee the reasonable dispersion of riches from capital or benefit producing exercises to whatever remains of society – in creating markets, as well as in develop ones.
- Inequality is reflected in the way that power, benefit, and impact are utilized as a part of the between linkages in supply chains (delineated by the Oxfam Behind the Brands battle).

What's more, there are empowering signals from worldwide organizations driving interest and consideration on riches balance.

- UN Sustainable Development Goal 10 decisively concentrates on decreasing disparity with particular targets, including continuously accomplishing and supporting salary development of the last 40% of the population at a rate higher than the national normal by 2030.
- The World Economic Forum additionally recognized riches disparity as a top pattern in 2015 and attempted a local investigation of the issue to better comprehend which imbalance related angles are needs in which districts (e.g., defilement, instruction, urbanization), which can better educate corporate comprehension of key markets. Implications for business

Oxfam sees different difficulties ahead for business identified with riches disparity, including:

- Radically expanding the share of significant worth caught by laborers and little scale makers.
- For example, accomplishing living wages for workers (and living earnings for little scale makers).
- Eliminating financial sexual orientation imbalance and sex segregation.
- Tackling the race-to-the-base on open administration to pull in speculation.

- Calling for a conclusion to the period of duty shelters.
- Breaking market fixation and tending to the unequal conveyance of energy
- For instance, new plans of action ought to be quickly settled to share esteem more impartially in supply chains.

26.2. EXAMPLES: BUSINESS ADDRESSING ISSUES RELATED TO WEALTH INEQUALITY

The cases of how organizations incorporating human rights-related data in their sustainability reports are:

- Ferrero shows expanding straightforwardness and broad announcing at the nation level (for key nations), exhibiting through revealing how business operations affect nearby economies. The report by nation on wage proportions (and kind of laborer), nearby administrative contracts and neighborhood provider spend.
- Mosaic presents giving an account of its worldwide group speculations – which are organized and arranged as identified with nourishment, water and nearby – making utilization of a microsite, Mosaic's GivingMap.

26.3. PRIORITIES AND NEXT STEPS

Need to explore and enhancing their providing details regarding riches imbalance:

- Start someplace – take a gander at your own operations and employees with a social rights point of view and after that consider extending the limits of this evaluation to incorporate the store network. Ask yourself: are there aberrations in the path in which the organization values individuals and exercises in the inventory network?
- Seek more data and detail on where the organization has made group level ventures and following going ahead.
- Consider how your organization as of now does, or can, impact access to fundamental items and administrations (e.g., sustenance, water, energy) on a worldwide premise or in the neighborhood groups in which you work.

- Explore effect and results-based estimation, regardless of the difficulties of middle people in supply chains, diverse purviews, distinctive exercises, and soon.
- Consider what your organization's most material effect identified with riches imbalance might be and create affect related KPIs.

CHAPTER 27

Data and Technology

27.1. CURRENT LANDSCAPE

Information and innovation are a piece of a steadily developing field of improvements. The CLG watches a speeding up of the measure of information being made accessible, incorporating into constant: more information has been made in the previous two years than in the whole past history of humankind. There are more methods of utilization of information crosswise over esteem chains, including systems of systems and a internet of things.

However, numerous corporate columnists indicate a requirement for more connection amongst information (and innovation) to comprehend impacts. For sure, for the world to meet SDG target 12.6 – to "empower organizations, particularly expansive and transnational organizations, to embrace sustainable practices and to coordinate sustainability data into their announcing cycle" – we should better comprehend and utilize information. The UN knows this and requires an information insurgency to make information bolster improvement and exhibiting why it truly matters. We know certainly that better accumulation of and access to information (frequently through innovation) empowers revealing that depends on increasingly and better information. Better information empowers partners to comprehend an organization's execution and thus settle on better choices.

Innovation can likewise empower partners to get to information more successfully. We watch corporate announcing utilizing information and innovation from multiple points of view: how organizations are straightforwardly exhibiting and imparting information remotely, how innovation can empower information openness and introduction, and how information can educate partner basic leadership.

27.2. DRIVERS OF CORPORATE EXECUTION ON INFORMATION AND INNOVATION

Pavan Sukhdev, Founder and CEO of GIST Advisory, joined the Lab on information and innovation as a specialist to extend individuals' reasoning on information and particularly what information organizations may as of now have on the off chance that they choose to esteem it, displaying a multi-capital esteem system. As far as he can tell with organizations, sparkling a light by gathering information on human, physical, regular and social capital can accomplish a multi-dimensional feeling of significant worth that is reflected in sustainability contemplations being better coordinated into organizations and announcing.

Understanding these perspectives better – and for a few, connecting a money-related an incentive to them – will empower organizations to recognize potential territories to enhance, to expand positive effects, diminish negative effects and at last increment business esteem.

27.3. SUGGESTIONS FOR BUSINESS

With regards to corporate detailing, information and innovation are regularly observed as an open door and a test in equivalent measure.

Challenges:

- Securing adequate inward purchase in.
- Promoting the way of life: making mindfulness for good utilization of the inward frameworks that convey high caliber, practically identical information.
- Availability of delicate and classified information.
- Need for more scientific instruments to better comprehend information.
- Manually ordering information crosswise over spreadsheets or decentralized frameworks, or transitioning to new, more imaginative frameworks to oversee information and data.

Openings:

- Online revealing as an inventive device to impart information adequately.

- Embedding sustainability information (counting gathering and auspicious accommodation) into targets and execution administration frameworks.
- Being proactive and keeping associates educated around the procedure and results.
- Monitor – and give criticism circles to information suppliers – both the positive and the negative effects of sustainability issues followed.
- Understand the flow and different requests on similar information to enhance the data channels and lower the weight for associates, where conceivable.

27.4. CASES: BUSINESS TENDING TO INFORMATION AND INNOVATION

Cases of how organizations exhibit information in their detailing, concentrating on internet announcing, intelligent revealing that utilization information to give the client a chance to pick how and what they read, infographics that present mind-boggling or a lot of information in a simple to peruse design, and regular execution refreshes:

- TIM presents its 2015 online report also outlined and takes into account simple route to its key concentrate on social advancement, natural assurance, and computerized culture and presents the report with a video.
- CLP opens up a few datasets to enable searchers to interface with and redo their perspective of CLP's information, and it covers money related, operational, natural, and social and administration execution information
- Grupo Nutresa presents complex information in a simple to peruse design by outwardly mapping the organization's exercises and commitments crosswise over many issues to particular SDGs.

27.5. NEEDS AND NEXT STRIDES

Needs to explore and enhancing their revealing using information and innovation:

- Use information and innovation to approach partners in various ways.
- Use innovation to facilitate the gathering of information by moving towards more advanced and brought together information frameworks.
- Explore and better comprehend the requirements of various partner gatherings and tailor the introduction of information in like manner.
- Present data to partners (e.g., infographics, report microsites) more successfully. Future Focus:
- Greater desires: Overall, organizations are encountering more prominent weight from partners on sustainability issues, as confirm by improvements, for example, expanding direction.
- Explore and test partner enthusiasm for revealing in various arrangements and frequencies and consider how announcing businesses may deal with this.
- Closely screen and investigate the four patterns (environmental change, human rights, riches disparity and information, and innovation), particularly the less developed regions of human rights and riches imbalance.
- Clearly convey the estimation of the development towards GRI Standards in connection to new or rising ranges: GRI can give more prominent clearness and direction on divulgences and guidelines however ought not to be prescriptive on the revealing procedure.
- Reset targets: Companies are resolved to return to their objectives; how might they be made longer-term and drive sustainable results? Will more effect and results-based estimation be actualized?
- Provide direction to organizations on targets. Would it be advisable for them to be science based? Setting-based? Speak to what's coming to the organization of a test or effect?
- Country-level and nearby focal points: There is a craving to search out more nation level and neighborhood information over these patterns – on wages, on striking human rights, to line up with government anticipates atmosphere, and soon.
- Explore promote direction on the most proficient method to report adequately on expense rehearse crosswise over various wards,

which governments can bolster this.

- Support inquiring about, benchmarking and direction on ranges, for example, least wages and living wages in various locales or human rights benchmarking to help organizations look at and report.
- Deepen the comprehension of effect: The inventory network is viewed as the following wilderness for some organizations and the need to better comprehend dangers and effects is foremost, regardless of whether that is investigating extension three discharges, leading human rights affect appraisals or investigating how riches imbalance plays out.
- Support revealing businesses' enthusiasm for effect estimation.
- Improve information: There are enhancements to be made to interior information accumulation procedures to enhance revealing against these patterns and bolster better understanding and dealing with the issues.
- Pursue further or more profound arrangement (e.g., GRI with different systems) so information is gathered once for various purposes.
- Encourage peer learning and sharing of best practice.
- SDGs matter: There is solid accord that the understanding of the Sustainable Development Goals (SDGs) is critical and will affect corporate sustainability motivation.

CHAPTER 28

Trends of Global Reporting

28.1. HISTORY OF REPORTING

1994–2000 – Birth of Reporting:

- First movers
- Origins of GRI

2000–2006 Era of Innovation and Rapid Growth:

- GRI G2 and G3
- AA1000AS

2006–2012 Era of Calm and Steady Growth:

- GRI only game in town
- Shared sense of best practice

2012–2015 Era of Disruption and Questioning:

- G4, SASB, IIRC
- Regulation and customer scrutiny 2015 – plus Era of digital technologies

28.2. THREE KEY GLOBAL REPORTING TRENDS

- **Materiality:** Companies are focusing on sustainability management and disclosure through reporting on topics of highest priority (and most relevance) to the company and stakeholders.
- **Integration:** Several companies are integrating sustainability management into the core of the business, and producing integrated reports as an output.

- **Interactive:** Companies are thinking about difference between report and reporting; a report is static; reporting is more dynamic: multi-channel, ongoing, and/or social.

28.3. THE DIFFERENT DIMENSIONS OF MATERIALITY

There are multiple criteria that can be used when applying the materiality principle and prioritizing sustainability issues. Here are some of the most common examples.

- Importance to business success;
- Importance to sustainability and/or stakeholders;
- Changing importance overtime;
- Ability of the company to influence;
- Extent of current management.

28.4. MATERIALITY ANALYSIS

Aims to identify the societal and environmental issues that present risks or opportunities to a company while taking into consideration the issues of most concern to external stakeholders. The process of labeling material topics in a company requires a comprehensive framework that methodologically identifies and prioritizes issues, risks, and opportunities.

28.5. HOW IT WORKS?

Define important areas/issues and internally organize these issues in order of importance. Ask stakeholders (contractors, suppliers, clients, and community) to organize the same issues in order of importance.

Use the results to create a Materiality Matrix that will order the issues according to the importance of all stakeholders confronted with the importance that the company has established.

Those issues that are found to be important for the company and for the total of other stakeholders (top right of the image) will be further developed into actions in order to maximize our positive impact and increase the added value provided.

Here you have an example of a materiality matrix (Figure 28.1):

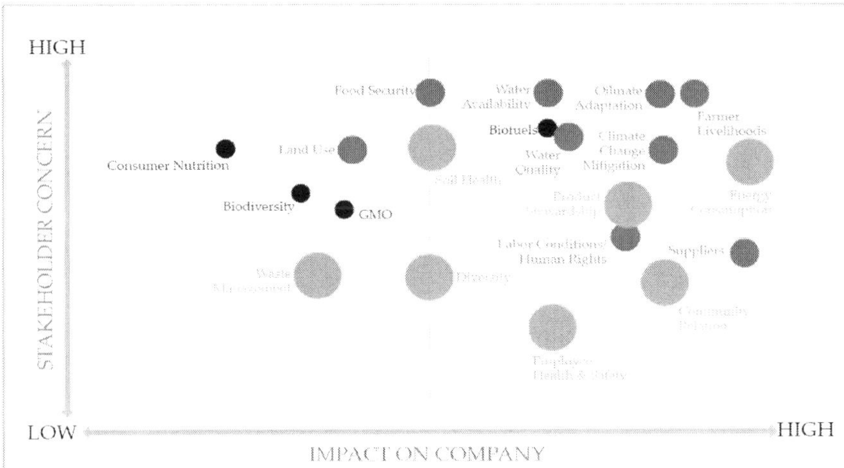

Figure 28.1: Materiality analysis example.

Source: ISO (2017)

28.6. FEATURES OF INTEGRATION

- Integrated thinking not combined reporting. An integrated report is not about just combining two previously separate publications into one; it is about weaving together a single, integrated narrative
- Single approach to materiality. An integrated report requires one single threshold for information considered important enough to include
- Entry point to more detailed information. Rather than pack in every detail, an integrated report should be logically structured with effective navigation and cross-referencing to more information available elsewhere
- Based on concepts of value creation and capitals. An integrated report explains how the organization interacts with the external environment and six capitals (financial, manufactured, intellectual, human, social and natural) to create value over the short, medium and long-term.

28.7. PATHWAYS TO INTEGRATED REPORTING

An integrated set of reports – not one report alone – meets the needs of diverse report users.

- Same look and feel: Have common branding, design, and format.
- Same theme: Highlight the same basic narrative, conveyed in a CEO letter.
- Same performance overview: Contain one performance dashboard, covering financial and sustainability.
- Same approval process: Follow identical review and sign-off protocols.

28.8. INTERACTIVITY

Thinking through the difference between a report and reporting:

- Some stakeholders want a "single moment in time" annual sustainability report, while others expect a depth of analysis that cannot be contained in a printed report. Companies are responding through the creation of a reporting system to satisfy multiple stakeholders, and allow for customization

28.9. LEVERAGING WEB 2.0

- Companies recognize the dynamic debates and insights are to be found on social media. Some are sharing their sustainability agenda and inviting contributions from stakeholders. Many are translating data into pictures, videos, snippets, blogs, contacts, and posts.

28.10. KEY TAKEAWAY

We're in the era of disruption (and innovation), where one size does not fit all – companies are rethinking reporting approaches, and forging new reporting trends which deliver value across multiple dimensions

CHAPTER 29

Certifications General Information

29.1. UNITED NATIONS (UN)

The UN Global Compact intends to prepare a worldwide development of feasible organizations and partners to make the world they need. That is their vision.

To get this going, the UN Global Compact backings organizations to:

- Do business dependably by adjusting their techniques and operations to 10 principles on human rights, work, condition and against defilement; and
- Take key activities to progress more extensive societal objectives, for example, the UN Sustainable Development Goals (SDGs), with an accentuation on cooperation and advancement.

29.1.1. United Nations (UN) Global Compact: 10 Principles

1. Human Rights:

a. Principle 1: Businesses should support and respect the protection of internationally proclaimed human rights; and

b. Principle 2: Make sure that they are not complicit in human rights abuses.

2. **Labor:**

c. **Principle 3:** Businesses should uphold the freedom of association and the effective recognition of the right to collective bargaining;

d. **Principle 4:** The elimination of all forms of forced and compulsory labor;

e. **Principle 5:** The effective abolition of child labor; and

f. **Principle 6:** The elimination of discrimination in respect of employment and occupation.

3. **Environment:**

g. **Principle 7:** Businesses should support a precautionary approach to environmental challenges;

h. **Principle 8:** Undertake initiatives to promote greater environmental responsibility;

i. **Principle 9:** Encourage the development and diffusion of environmentally friendly technologies.

4. **Anti-Corruption:**

j. **Principle 10:** Businesses should work against corruption in all its forms, including extortion and bribery.

5. **Results (Figure 29.1):**

9,269	166	43,142
COMPANIES	COUNTRIES	PUBLIC REPORTS

Figure 29.1: UNGC certifications figures.

Source: UNGC (2017).

29.1.2. Sustainable Development Goals (SDGs): 17 Goals to Transform Our World

Governments, businesses, and civil society together with the United Nations (UN) have started to mobilize efforts to achieve the Sustainable Development Agenda by 2030. Universal, inclusive, and indivisible, the Agenda calls for action by all countries to improve the lives of people everywhere.

In 2015, countries adopted the 2030 Agenda for Sustainable Development and its 17 SDGs. In 2016, the Paris Agreement on climate change entered into force, addressing the need to limit the rise of global temperatures (Figure 29.2).

Figure 29.2: UNGC sustainable development goals.

Source: UNGC (2017).

CHAPTER 30

ISO 26000

30.1. WHAT ADVANTAGES CAN BE ACCOMPLISHED BY UPDATING ISO 26000?

An association's execution on social obligation can impact, in addition to other things:

- Competitive preferred standpoint.
- Reputation.
- The capacity to draw in and hold laborers or individuals, clients, customers and clients.
- The upkeep of worker resolve, duty, and profitability.
- The impression of speculators, proprietors, givers, supports and the money related group.
- Relationships with organizations, governments, the media, providers, companions, clients, and the group in which it works.

30.2. WHO CAN PROFIT BY ISO 26000 AND HOW?

ISO 26000 gives direction to a wide range of association, paying little respect to their size or area, on:

- Ideas, terms, and definitions identified with social duty.
- Foundation, patterns, and qualities of social duty.
- Standards and works on identifying with social duty.
- Center subjects and issues of social duty.
- Coordinating, executing and advancing socially capable conduct all through the association and, through its approaches and

practices, inside its range of prominence.

- Recognizing and drawing in with partners.
- Conveying duties, execution and other data identified with social obligation.

ISO 26000 is proposed to help businesses in adding to sustainable improvement. It urges them to go past legitimate consistence, perceiving that consistence with the law is a key obligation of any association and a fundamental piece of their social duty program.

The standard tries to advance a typical comprehension of social obligation while supplementing – however not supplanting – other existing instruments and activities. While applying ISO 26000, businesses ought to consider societal, natural, lawful, social, political and hierarchical assorted qualities and contrast in monetary conditions, while being predictable with global standards of conduct.

Not for confirmation ISO 26000 is not an administration framework standard.

It doesn't contain necessities and, all things considered, can't be utilized for confirmation. Any offer to affirm, or claims to be ensured, against ISO 26000 would be a deception of its goal and reason.

What specifically does ISO 26000 contain? ISO 26000 addresses seven core subjects of social responsibility defined in the standard and portrayed in the following graphic (Figure 30.1).

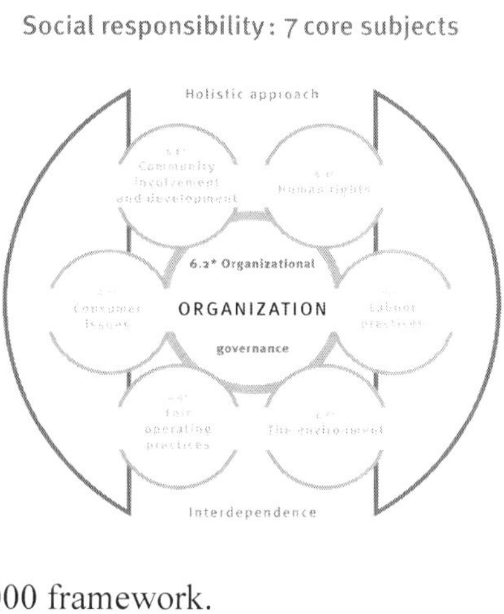

Figure 30.1: ISO 26000 framework.

Source: ISO (2017)

CHAPTER 31

GRI

GRI is a global free association that helps organizations, governments, and different businesses comprehend and impart the effect of business on basic sustainability issues, for example, environmental change, human rights, defilement and numerous others.

Their vision is to make a future where sustainability is essential to each association's basic leadership prepare.

Their central goal is to engage leaders all over the place, through our sustainability models and multi-partner system, to make a move towards a more economical economy and world.

1. **They Accept:**
   - In the energy of a multi-partner handle and comprehensive system.
   - Transparency is an impetus for change.
   - Our measures engage educated basic leadership.
   - A worldwide point of view is expected to change the world.
   - Public intrigue ought to drive each choice an association makes.
2. **Vital Needs:** GRI is based upon a novel multi-partner rule which guarantees the interest and aptitude of assorted partners in the improvement of its gauges. The advantage of this expansive and comprehensive approach is that businesses can have trust in the vigor of partner cooperation while considering an extensive variety of sustainability issues. GRI is seeing a more prominent interest for sustainability data for particular uses exhibited by the development of new announcing activities. Despite the fact that the utilization of the data varies, the inspiration for giving it is regularly the same: to assemble trust (Figure 31.1).

Figure 31.1: GRI framework.

Source: GRI (2017)

31.1. EMPOWERING SMART POLICY

GRI's worldwide group traverses a large number of columnists yet our degree stretches out past topography and into different fields of work. They work with governments, worldwide businesses, and capital markets. Thus, of this work, 27 nations utilize GRI in their sustainability approaches and seek us for direction as the universes most broadly utilized sustainability detailing benchmarks. Furthermore, they have long-standing joint efforts with more than 20 universal businesses, for example, the UNGC, OECD, and the UN Working Group on Business and Human Rights.

GRI teams up with different businesses on various worldwide driving edge activities which concentrate on handling a portion of the world's most squeezing difficulties, for example, environmental change, human rights, and defilement. Through this sort of joint effort, they are finding imaginative methods for utilizing sustainability detailing data to help bolster organizations specifically.

Consistently they help developed nations put millions in sustainable improvement exercises in developing markets to address neediness related issues, for example, water utilize, arrive utilize and sexual orientation correspondence. GRI's Standards keep on playing an imperative part in these sorts of speculation streams guaranteeing that sustainability issues are mulled over by these businesses as they improve their aggressive edge in the neighborhood and global markets.

31.2. MORE REPORTERS, BETTER REPORTING

GRI is best known for driving the improvement of the sustainability revealing procedure and keeps on building their worldwide group, which incorporates a great many prepared professionals and columnists. Among this differing group, they additionally have many center individuals who work straightforwardly with them to champion their central goal.

As a worldwide autonomous association, the provincial centers empower them to address novel issues crosswise over topographies and set up a genuinely worldwide nearness. GRI works with a developing group to create industry-driving exercises to promote the estimation of sustainability detailing. They support and empower those businesses that don't right now answer to do as such, particularly advancing the take- up of revealing among little and medium-sized ventures.

GRI persuades and underpins those that as of now do answer to make progress toward a superior nature of revealing through our preparation and information sharing exercises. What's more, for those that is driving the way, they work in coordinated effort to investigate and imagine enormous picture difficulties, for example, the eventual fate of revealing through their corporate initiative engagement exercises.

GRI is focused on giving worldwide models to sustainability detailing and keep on strengthening their believability and part as a standard-setter through late changes to administration structure. This incorporates the production of a different administration body to create and support GRI's Sustainability Reporting Standards to expand the autonomy of these exercises. Guaranteeing that more organizations and governments' esteem sustainability data will remain an essential piece of GRIs DNA, however, the time has desired them to expand on this inheritance and catalyze another time where organizations and governments settle on vital choices in light of these vital sustainability issues.

31.3. MOVING BEYOND REPORTS

GRIs concentrate has dependably been on the detailing procedure and the estimation of the data that originates from it. While the sustainability report remains a vital yield of the detailing procedure, they should now move past the report itself to guarantee that chiefs have entry to the high caliber and solid data they are progressively requesting.

From the utilization of sustainability data in business insight to the real formation of new lines of business, unmistakably this data isn't only for sustainability experts. It is for financial specialists to empower more prominent trust in their ventures, clients in their items and administrations, subjects in their legislatures et cetera.

These gatherings and people all need distinctive data, and need to get to it in various ways. Today we are seeing a rush of new routes in which data from the sustainability detailing procedure is being utilized by differing gatherings of partners to address their particular needs.

Yet, for this data to genuinely enable economical choices in each association, it must be more open, equivalent and accessible progressively. GRIs part in deduction past the sustainability report is to help the inexorably various scope of clients take advantage of the sustainability data they require, empowering them to utilize this data in more imaginative ways.

31.4. ADVANCEMENT AND COLLABORATION

The sustainability scene is continually changing and GRI has dependably trusted in the requirement for development. GRI's Standards are a free, open great which have effectively motivated the making of new systems, businesses, arrangements, and professions all centered around enhancing the estimation of sustainability data.

GRI keeps on driving the exchange on incorporating sustainability and monetary data, working with various businesses and partners to further this plan. Effective development requires another soul of joint effort and comprehensiveness in the announcing field, guaranteeing businesses cooperate towards a shared objective of building trust in organizations and governments. A portion of the best developments, be that as it may, will originate from coordinated efforts past the detailing field.

This is the place innovation and enormous information have a gigantic part to play and we're working with the world's driving innovation organizations to move to a future 'past reports.' This will help all partners get to the data they have to settle on better business and approach choices, in light of the world's most generally utilized and trusted sustainability detailing benchmarks.

31.5. GLOBAL REPORTING INITIATIVE (GRI)

- Increasing understanding of the connection between ESG issues and sustainable business value.
- Increasing interest in organizational focus – what and where.

Bibliography

1. Ahlering, J., (2009). "*RE: 2009 Ceres–ACCA Sustainability Reporting Awards*." 10–15.
2. Alcan, (2010). *Alcan Composites: Serving Customers with Multi-Material Solutions*. Retrieved from: http://www.alcancomposites.com/contact.html (Accessed on 8 July 2019).
3. Arrow, K. J., Chenery, H. B., Minhas, B. C., & Solow, R. M., (1961). Capital-labor substitution and economic efficiency. *The Review of Economics and Statistics, 43*(3), 225–250.
4. As You Sow, (2009). *Proxy Preview*. Spring – 2009. Retrieved from: http://www.asyousow.org/csr/proxyvoting.shtml.
5. Asif, M., Bruijn, E. J. D., Fisscher, O. A. M., & Searcy, C., (2010). Meta-management of integration of management systems. *The TQM Journal, 22*(6), 570–582.
6. Asif, M., Searcy, C., & Fisscher, O. A. M., (2010). Meta-management of corporate social responsibility. In: Lindgreen, A., Sen, S., Maon, F., & Vanhamme, J., (eds.), *Sustainable Value Chain Management: Analyzing, Designing, Implementing, and Monitoring for Social and Environmental Responsibility* (25–40). Gower Publishing.
7. Auger, P., & Devinney, T., (2007). Do what consumers say matter? The misalignment of preferences with unconstrained ethical intentions. *Journal of Business Ethics, 76*(4), 361–383.
8. Auger, P., Devinney, T. M., Louviere, J. J., & Burke, P. F., (2008). Do social product features have value to consumers? *International Journal of Research in Marketing, 25*(3), 183–191.

9. Auger, P., Devinney, T. M., Louviere, J. J., & Burke, P. F., (2010). The importance of social product attributes in consumer purchasing decisions: A multi-country comparative study. *International Business Review, 19*(2), 140–159.
10. Aupperle, K. E., (1984). *An Empirical Measure of Corporate Social Orientation*. Research in corporate social policy and performance. Greenwich: JAI Press, Google Scholar.
11. Aupperle, K. E., Carroll, A. B., & Hatfield, J. D., (1985). An empirical examination of the relationships between corporate social responsibility and profitability. *Academy of Management Journal, 28*(2), 446–463. View Article Google Scholar.
12. Autodesk, (2009). "*Improving Environmental Performance Through Life Cycle Assessment and Digital Prototyping*." Autodesk.com. Retrieved from: http://usa.autodesk.com/company/sustainable-design/articles/improving-environmental-performance-through-life-cycle-assessmentand-digital-prototyping (Accessed on 8 July 2009).
13. Awaysheh, A., & Klassen, R. D., (2010). The impact of supply chain structure on the use of supplier socially responsible practices. *International Journal of Operations and Production Management, 30*(12), 1246–1268.
14. Bagozzi, R. P., & Yi, Y., (1988). On the evaluation of structural equation models. *Journal of the Academy of Marketing Science, 16*(1), 74–94.
15. Bagozzi, R. P., Gopinath, M., & Nyer, P. U., (1999). The role of emotions in marketing. *Journal of the Academy of Marketing Science, 27*(2), 184–206.
16. Bai, C., & Sarkis, J., (2010). Green supplier development: Analytical evaluation using rough set theory. *Journal of Cleaner Production, 18*(12), 1200–1210.
17. Bavly, D., (1999). *Corporate Governance and Accountability: What Role for the Regulator, Director, and Auditor? Quorum Books* (pp. 10–25). Westport, Conn, London.
18. Baxter, (2009). "*Baxter Facilities Worldwide Observe 'Baxter World Environment Week' With Local Initiatives*." Baxter.com. Retrieved from: http://www.baxter.com/about_baxter/press_ (Accessed on 8 July 2019).
19. BBMG, (2008). *Bearing Point Management and Technology Consultants*. Insight Survey Report: 2008 Supply chain monitor

"how mature is the green supply chain?" Retrieved from: http://www.escp-eap.eu/ uploads/media/Supply_Chain_Observatory_2008.pdf (Accessed on 8 July 2019).

20. BBMG, (2009). *BBMG Conscious Consumer Report: Redefining Value in a New Economy*. Retrieved from: http://www.bbmg.com/enewsletter/bbmg_be_0409.html (Accessed on 8 July 2019).
21. Becker-Olsen, K. L., Cudmore, B. A., & Hill, R. P., (2006). The impact of perceived corporate social responsibility on consumer behavior. *Journal of Business Research, 59*(1), 46–53.
22. Bell, S. T., Kuriloff, P. J., & Lottes, I., (1994). Understanding attributions of blame in stranger rape and date rape situations: An examination of gender, race, identification, and students' social perceptions of rape victims. *Journal of Applied Social Psychology, 24*(19), 1719–1734.
23. Bendoly, E., Donohue, K., & Schultz, K. L., (2006). Behavior in operations management: Assessing recent findings and revisiting old assumptions. *Journal of Operations Management, 24*(6), 737–752.
24. Berman, S. L., Wicks, A. C., Kotha, S., & Jones, T. M., (1999). Does stakeholder orientation matter? The relationship between stakeholder management models and firm financial performance. *Academy of Management Journal, 42*(05), 488–506.
25. Blackorby, C., & Donaldson, D., (1982). Ratio-scale and translation-scale full interpersonal comparability without domain restrictions: Admissible social-evaluation functions. *International Economic Review, 23*(2), 249–268.
26. Blanchard, D., (2008). "*SC Johnson Finds a 'Greener' Way to Load Trucks*." Industry Week. Retrieved from: http://www. industryweek.com/articles/sc_johnson_finds_a_greener_way_to_load_ (Accessed on 8 July 2019).
27. Bordoff, J., & Pascal, J. N., (2008). *Pay-As-You-Drive Auto Insurance: A Simple Way to Reduce Driving-Related Harms and Increase Equity*. The Brookings Institution. Retrieved from: http://www.brookings.edu/papers/2008/07_payd_bordoffnoel.aspx (Accessed on 8 July 2019).
28. Boubaker, S., & Nguyen, D. K., (2012). *Board Directors and Corporate Social Responsibility* (pp. 50–66). Palgrave Macmillan M.U.A., GB.
29. Bowen, H., (1953). *Social Responsibilities of the Businessman*. New York: Harper and Row. Google Scholar.
30. Braunsberger, K., & Buckler, B., (2011). What motivates consumers

to participate in boycotts: Lessons from the ongoing Canadian seafood boycott. *Journal of Business Research, 64*(1), 96–102.

31. Brickman, C., & Drew, U., (2008). *Climate Change and Supply Chain Management.* McKinsey Quarterly. Retrieved from: https://www.mckinseyquarterly.com/ghost.aspx?ID=/Climate_change_and_supply-chain_management_2175 (Accessed on 8 July 2019).
32. Brown, K. A., Willis, P. G., & Prussia, G. E., (2000). Predicting safe employee behavior in the steel industry: Development and test of a sociotechnical model. *Journal of Operations Management, 18*(4), 445–465.
33. Bruel, O., Olivier, M., & Pierre-Francois, T., (2009). *Sustainable Procurement: A Crucial Lever to End the Crisis?* HEC Paris, SNCF, and Eco Vadis. 2009. Retrieved from: http://www.hec.Edu/News/Homepage/Knowledge-impact/The-HEC-EcoVadis–2009-Barometer-Sustainable-Procurement-a-crucial-lever-to-end-the-crisis (Accessed on 8 July 2019).
34. BSR, (2017). *Moving the Needle: Protecting the Rights of Garment Factory Workers.* Retrieved from: http://www.bsr.org/ reports/BSR_LeviStraussFoundation2009.pdf (Accessed on 8 July 2019).
35. Bunn and Simon, (2009). Greenhouse gas reduction as an additional benefit of optimal pump scheduling for water utilities. *IBM Journal of Research and Development.* Retrieved from: http://www.derceto.com/cms/lib/334.pdf (Accessed on 8 July 2019).
36. Burke, L., & Logsdon, J., (1996). How corporate social responsibility pays off. *Long Range Planning, 29*(4), 495–502.
37. Burton, B. K., & Hegarty, W. H., (1999). Some determinants of student corporate social responsibility orientation. *Business and Society, 38*(2), 188–2005 (View Article Google Scholar).
38. California Energy Commission, (2005). *California's Water-Energy Relationship.* Prepared in support of the 2005 integrated energy policy report proceeding. Retrieved from: http://www.energy.ca.gov/2005publications/CEC–700–2005–011/CEC–700–2005–011-SF.PDF (Accessed on 8 July 2019).
39. Calvert Investments, (2008). *Examine the Cracks in the Ceiling: A Survey of Corporate Diversity Practices in the Calvert Social Index.* Retrieved from: http://www.calvertgroup.com/NRC/Literature/Documents/CorporateDiversity2008.pdf (Accessed on 8 July 2019).

40. Calvert Investments, (2010). *"The Calvert Social Index." Calvert Investments, a Product of UNIVI.* Retrieved from: http://www.calvertgroup.com/sri-index.html (Accessed on 8 July 2019).

41. Campbell, J. L., (2007). Why would corporations behave in socially responsible ways? An institutional theory of corporate social responsibility. *Academy of Management Journal, 32*(3), 946–967.

42. Carroll, A. B., (1979). A three-dimensional conceptual model of corporate social performance. *Academy of Management Review, 4*, 497–505 (Google Scholar).

43. Carroll, A. B., (1991). The pyramid of corporate social responsibility: Toward the moral management of organizational stakeholders. *Business Horizons, 34*(4), 39–48 (View Article Google Scholar).

44. Carroll, A. B., (1999). Corporate social responsibility: Evolution of a definitional construct. *Business and Society, 38*, 268–295 (View Article Google Scholar).

45. Carroll, A. B., (2015a). Corporate social responsibility: The centerpiece of competing and complementary frameworks. *Organizational Dynamics, 44*, 87–96 (View Article Google Scholar).

46. Carroll, A. B., (2015b). The state of CSR in 2015 and beyond. *Global Compact International Yearbook* (pp. 10–13). Macondo Publishers, Global Logistics Partner, Deutsche Post DHL Group.

47. Carroll, A. B., & Buchholtz, A. K., (2015). *Business and Society: Ethics, Sustainability and Stakeholder Management* (9th edn., pp. 25–30). Stamford: Cengage Learning. Google Scholar.

48. Carroll, A. B., & Shabana, K. M., (2010). The business cases for corporate social responsibility: A review of concepts, research, and practice. *International Journal of Management Reviews, 12*(1), 85–105, (View Article Google Scholar).

49. Carroll, A. B., Lipartito, K. J., Post, J. E., Werhane, P. H., & Goodpaster, K. E., (2012). *Corporate Responsibility: The American Experience*. Cambridge: Cambridge University Press. Google Scholar.

50. Carter, C. R., (2000). Ethical issues in international buyer-supplier relationships. *International Journal of Physical Distribution & Logistics Management* (Vol. 45), 140–160.

51. Carter, Craig & Jennings, Marianne, (2004). The role of purchasing in corporate social responsibility: A structural equation analysis. *Journal of Business Logistics, 25*, 145–186.

52. Cassidy, M., (2014). *Assessing Gaps in Indicator Availability and Coverage* (pp. 100–150). SDSN working paper. New York: Sustainable development solutions Network Conservation international, 2015. Ocean Health Index. Washington D. C.: Conservation International.
53. Castka, P., Bamber, C. J., Bamber, D. J., & Sharp, J. M., (2004). Integrating corporate social responsibility (CSR) into ISO management systems – in search of a feasible CSR management system framework. *The TQM Magazine, 16*(3), 216–224.
54. Ceres, (2017). "*Bloomberg L. P. Joins the Ceres Network.*" Ceres.com. http://www.ceres.org/Page.aspx?pid=1067 (Accessed on 8 July 2019).
55. Ceres, (2017). "*Gap, eBay, and Symantec Join Business Coalition Calling for Action on Clean Energy and Climate Change.*" Ceres.com. Boston, MA. Retrieved from: http://www.ceres.org/Page.aspx?pid=1060 (Accessed on 8 July 2019).
56. Ceres, (2017). "*Identifying Stakeholders.*" Diagram. Boston, MA: 2009.
57. Ceres, (2017). "*Sustainability Shareholder Resolution Trends 2000–2009.*" Graph. Boston, MA: 2009.
58. Chrisman, J. J., & Carroll, A. B., (1984). SMR forum: Corporate responsibility-reconciling economic and social goals. *Sloan Management Review* (pp. 59–65). Winter.
59. Ciliberti, F., Pontrandolfo, P., & Scozzi, B., (2008). Investigating corporate social responsibility in supply chains: A SME perspective. *Journal of Cleaner Production, 16*(15), 1579–1588.
60. Cisco, (2017). "*Cisco Study Finds Telecommuting Significantly Increases Employee Productivity, Work-Life Flexibility, and Job Satisfaction.*" cisco.com. San Jose, California. Retrieved from: http://newsroom.cisco.com/dlls/2009/prod_062609.html (Accessed on 8 July 2019).
61. Clarkson, M. B. E., (1995). A stakeholder framework for analyzing and evaluating corporate social performance. *Academy of Management Review, 20*, 92–117 (Google Scholar).
62. Clif, B., (2017). "*Moving Towards Sustainability: Working to Reduce Our Ecological Footprint* (Issue 4)." Cliff Bar sustainability newsletter, Winter. Retrieved from: http://www.clifbar.com/uploads/default/Sustainability_Issue4.winter2006.pdf (Accessed on 8 July 2019).
63. Cone, (2009). *Consumer Environmental Survey Fact Sheet*, 18.

64. Cornett, M. M., Marcus, A. J., Saunders, A., & Tehranian, H., (2007). "The impact of institutional ownership on corporate operating performance." *Journal of Banking and Finance, 31*(6), pp. 1771–1794.

65. Craig R. C., Dale, S. R., (2008) "A framework of sustainable supply chain management: Moving toward new theory." *International Journal of Physical Distribution & Logistics Management, 38(5), 360–387.*

66. Cramer, J., (2008). Organizing corporate social responsibility in international product chains. *Journal of Cleaner Production, 16*(3), 395–400.

67. Crane, A., & Matten, D., (2007). *Business Ethics: Managing Corporate Citizenship and Sustainability in the Age of Globalization* (2nd edn.). Oxford University Press.

68. Crane, A., Matten, D., & Spence, L., (2008). *Corporate Social Responsibility: Readings and Cases in a Global Context* (2nd edn., pp. 3–26). New York: Routledge (Google Scholar).

69. Crossley, C. D., (2009). Emotional and behavioral reactions to social undermining: A closer look at perceived offender motives. *Organizational Behavior and Human*, 48–60.

70. Crowne, D. P., & Marlowe, D., (1960). A new scale of social desirability independent of psychopathology. *Journal of Consulting Psychology, 24*(4), 349–354.

71. Daft, R. L., (2008). *The New Era of Management.* Thomson: South-Western, Mason: OH.

72. Dahlsrud, A., (2006). How corporate social responsibility is defined: An analysis of 37 definitions. *Corporate Social Responsibility and Environmental Management, 15*, 1–13 (View Article Google Scholar).

73. Dahlsrud, A., (2008). How corporate social responsibility is defined: An analysis of 37 definitions. *Corporate Social Responsibility and Environmental Management, 15*(1), 1–13.

74. Dale, B. G., (2003). *Managing Quality* (pp. 1–40). Blackwell Publishing, Oxford.

75. Davenport, K., (2000). Corporate citizenship: A stakeholder approach for defining corporate social performance and identifying measures for assessing it. *Business and Society, 20*(2), 210–219.

76. Davis, K., (1975). Blamestorm. *Business and Society: Environment and Responsibility* (3rd edn.). New York: McGraw-Hill Book Company (Google Scholar).

77. Davis, S. C., Susan, W. D., & Robert, G. B., (2009). *Transportation Energy Data Book* (28th edn.), U.S. Department of Energy. Retrieved at: http://www-cta.ornl.gov/data/index.shtml (Accessed on 8 July 2019).
78. Dawar, N., & Pillutla, M. M., (2000). Impact of product-harm crises on brand equity: The moderating role of consumer expectations. *Journal of Marketing Research, 37*, 27–59.
79. De Colle, S., & Gonella, C., (2002). The social and ethical alchemy: An integrative approach to social and ethical accountability. *Business Ethics: A European Review, 11*(1), 86–96.
80. De Colle, S., & Gonella, C., (2003). Corporate social responsibility: The need for an integrated management framework. *International Journal of Business Performance Management, 5*(2/3), 199–212.
81. DeJoy, D. M., (1994). Managing safety in the workplace: An attribution. *Theory Analysis*, 100–150.
82. Delmas, M., & Montiel, I., (2009). *Greening the Supply Chain: When is Customer Pressure*, 35–80.
83. Deming, W. E., (1994). *The New Economics for Industry, Government, Education*. MIT Press, Cambridge, MA.
84. Deutsch & Claudia, (2017)."*Companies Giving Green an Office*." The New York Times. Retrieved from: http://www.nytimes.com/2007/07/03/business/03sustain.html (Accessed on 8 July 2019).
85. Doerr & John, (2017). "*Green Beat 2009: Intro and Opening Keynote*." In green beat 2009: The conference on the smart grid. Retrieved from: http://events.venturebeat.com/greenbeat2009// video/ (Accessed on 8 July 2019).
86. Eaton, J., Czyznielewski, A., & Dupuis, R., (2005). Judging up the corporate ladder: Understanding the social conduct of workers. *Journal of Applied Social,* 10–35
87. Eckerd, S., Hill, J., Boyer, K. K., Donohue, K., & Ward, P. T., (2013). The relative impact of attribute, severity, and timing of psychological contract breach on behavioral and attitudinal outcomes. *Journal of Operations Management, 31*(7/8), 567–578.
88. Ecobilan, (2017). "*Environmental Impact Case Studies for Buildings*." Retrieved from: http://fdes.ecobilan.com/fr/Fichede-declaration-environnementale-et-sanitaire-Achats-verts.php (Accessed on 8 July 2019).
89. Economist, (2008). Corporate social responsibility – the next question.

The Economist Special Report, 386(17), 8–10.

90. Economistonomist Intelligence Unit, (2006). *Doing Good: Business and the Sustainability Challenge*. EIU.com. Retrieved from: http://www.eiu.com/site_info.asp?info_name=corporate_sustainability (Accessed on 8 July 2019).
91. Economy, E., (2013). *China's Water Pollution Crisis*. http://thediplomat.com/2013/01/ (Accessed on 8 July 2019).
92. Edgeman, R. L., & Hensler, D. A., (2001). The AO chronicle: earth@omega or sustainability@alpha? *The TQM Magazine, 13*(2), 83–90.
93. Edmondson, V. C., & Carroll, A. B., (1999). Giving back: An examination of the philanthropic motivations, orientations, and activities of large black-owned businesses. *Journal of Business Ethics, 19*, 171–179 (View Article Google Scholar).
94. Elkington, J., (1999). *Cannibals with Forks: The Triple Bottom Line of 21st Century Business* (pp. 120–140). Capstone Publishing Ltd, Oxford.
95. Elkington, J., (2004) *Enter the Triple Bottom Line*. http://www.johnelkington.com/archive/TBL-elkington-chapter.pdf (Accessed on 8 July 2019).
96. Ellen, P. S., Webb, D. J., & Mohr, L. A., (2006). Building corporate businesses: Consumer attributions for corporate socially responsible programs. *Journal of the Academy,* 30–70.
97. Elsevier Journals, (2017). Most downloaded business horizons articles. http://www.journals.elsevier.com/business-horizons/most-downloaded-articles/ (Accessed on 8 July 2019).
98. Environmental Defense Fund, (2009). *Corporate Fleet Emissions Survey Report*. Retrieved at http://www.edf.org/documents/10434_fleet-benchmarking-survey.pdf (Accessed on 8 July 2019).
99. Ernst & Young, L. L. P., (2017). *Role of HR in Driving Sustainable Business Practices* (01–40). Published in India.
100. EurActive.com., (2017). "*HSBC: World Climate Business Revenue $2 Trin by 2020*." EurActiv.com. World business council for sustainable development. Retrieved from: http://www.euractiv.com/en/climate-change/hsbc-world-climate-businessrevenue–2-trln–2020/article–185596 (Accessed on 8 July 2019).
101. Eurosif., (2017). *Investment Consultants and Responsible Investment Study, 2009*. Received from: http://www.eurosif.org/media/files/eurosif_ investment_ consultants _restudy (Accessed on 8 July 2019).

102. Field, A., & Hole, G., (2003). *How to Design and Report Experiments* (pp. 38–80). Sage, London, UK.
103. Fincham, F., & Jaspars, J. M., (1980). *Attribution of Responsibility: From Man the Scientist to Man as Lawyer*. 10.1016/S0065–2601(08)60131–8.
104. Fitzpatrick & Kalpana, (2017). "*CalPERS Boosts Corporate Governance Strategy*." Pensions Gym. Retrieved from: http://www.pensionsgym.com/pensions-news/archive/more.asp?article=2926 (Accessed on 8 July 2019).
105. Foerstl, K., Reuter, C., Hartmann, E., & Blome, C., (2010). *Managing Supplier Sustainability*, 15–45.
106. Foley, K. J., (2005). *Meta Management: A Stakeholder/Quality Management Approach to Whole-of-Enterprise Management* (pp. 59–140). Standards Australia Ltd, Sydney.
107. Folkes, V. S., (1984). Consumer reactions to product failure: An attributional approach. *Journal of Consumer Research, 10*(4), 398–409.
108. Fombrun, C., Gardberg, N., & Barnett, M., (2000). Opportunity platforms and safety nets: Corporate citizenship and reputational risk. *Business and Society Review, 105*(1), 85–106.
109. Fragale, A. R., Rosen, B., Xu, C., & Merideth, I., (2009). The higher they are, the harder they fall: The effects of wrongdoer status on observer punishment recommendations and intentionality attributions. *Organizational Behavior and Human Decision Processes, 108*(1), 53–65.
110. Frantz, P., & Instefjord, N., (2007). "Socially and privately optimal shareholder activism." *Journal of Management and Governance, 11*(1), pp. 23–43.
111. Freeman, R. E., (1984). *Strategic Management: A Stakeholder Approach* (100–140). Pitman, Boston.
112. Frey, C. H., (2017). "*Identification and Evaluation of Potential Best Practices for Greenhouse Gas Reductions in Freight Transportation*." California air resources board chair's air pollution seminar series. Retrieved from: http://www.arb.ca.gov/research/ (Accessed on 8 July 2019).
113. Friedman, M., (1962). *Capitalism and Freedom*. Chicago: University of Chicago Press. (Google Scholar).

114. Friedman, M., (1985). Consumer boycotts in the United States, 1970–1980: Contemporary events in historical perspective. *Journal of Consumer Affairs, 19*(1), 96.

115. Frooman, J., (1999). Stakeholder influence strategies. *Academy of Management Review, 24*(2), 191–205.

116. Gallup, (2015). *Gallup World Poll* (pp. 25–35). Washington, D.C.: Gallup.

117. Gao, H., Knight, J. G., Zhang, H., Mather, D., & Tan, L. P., (2012). Consumer scapegoating during a systemic product-harm crisis. *Journal Marketing Management, 28*(11/12), 1270–1290.

118. Gao, J., & Bansal, P., (2013). Instrumental and integrative logics in business sustainability. *Journal of Business Ethics, 112*(2), 241–255.

119. Gap, (2017). "*Supply Chain: Our Program in Action*." Gap, Inc. 2007–2008, Social responsibility report 2009. http://www.gapinc.com/GapIncSubSites/csr/Goals/SupplyChain/Program/SC_Purchasing_Practices_Program.shtml (Accessed on 8 July 2019).

120. Garret-Peltier, H., James, H., & Robert, P., (2017). *The Economic Benefits of Investing in Clean Energy: How the Economic Stimulus Program and New Legislation Can Boost U.S. Economic Growth and Employment*. Amherst, Massachusetts. Retrieved from: http://www.americanprogress.org issues/2009/06/pdf/peri_report.pdf (Accessed on 8 July 2019).

121. Gassert, F., Reig, P., Luo, T., & Maddocks, A., (2013). *Aqueduct Country and River Basin Rankings: A Weighted Aggregation of Spatially Distinct Hydrological Indicators*. Working paper. Washington, D. C.: World Resources Institute (Accessed on 8 July 2019).

122. Gavronski, I., Klassen, R. D., Vachon, S., & Nascimento, L. F. M. D., (2011). A resource-based view of green supply management. *Transportation Research Part E: Logistics and Transportation Review, 47*(6), 872–885.

123. General, M., (2009). *General Mills Corporate Social Responsibility Report*. Retrieved from: http://www.generalmills.com/ corporate/commitment/NEW_CSR_2009.pdf (Accessed on 8 July 2019).

124. Godfrey, P. C., Merrill, C. B., & Hansen, J., (2009). The relationship between corporate social responsibility and shareholder value: An empirical test of the risk management hypothesis. *Strategic Management Journal, 30*(4), 425–445.

125. Goel, R., (2005). *Guide to Instruments of Corporate Responsibility: An Overview of 16 Key Tools for Labor Fund Trustees*. Schulich Business School, York University, Toronto.

126. Goleman and Daniel, (2017). "*Winning in an Age of Radical Transparency*." Harvard business publishing blog. Retrieved from: http://blogs.harvardbusiness.org/leadinggreen/2009/05/radicaltransparency.html (Accessed on 8 July 2019).

127. Green Buildings, Department of Buildings, (2017). "*Green Buildings*." Retrieved from: http://egov.cityofchicago.org/city/webportal/portalEntityHomeAction.do?entityName=Green+Buildings&entityN am eEnumValue=19 in January 2010 (Accessed on 8 July 2019).

128. Greener Design Staff, (2017). "*Canon Unveils First Products Developed with Life Cycle Thinking*." Greenbiz.com. Retrieved from: http://www.greenbiz.com/news/2009/09/23/canon-unveils-firstproducts-developed-life-cycle-thinking (Accessed on 8 July 2019).

129. Greenpeace International, (2017). *Eating Up the Amazon*. Retrieved from: http://www.greenpeace.org/international/press/reports/ eating-up-the-amazon (Accessed on 8 July 2019).

130. Greenpeace, (2009). *Playing Dirty: How Factories Hide Their Pollution in China,* 1–10.

131. GRI, (2006). *Sustainability Reporting Guidelines: Version 3.0* (pp. 20–45). GRI, Amsterdam, the Netherlands.

132. GRI, (2017). *Global Reporting Initiative*. "Number of companies worldwide reporting on their sustainability performance reaches record high, yet still a minority." Retrieved at: http://www.globalreporting.org/NewsEventsPress/LatestPressReleases/2009/PressRelease_14_July_2006_1000GRIReports.htm (Accessed on 8 July 2019).

133. GRI, (2017). *Global Reporting Initiative, Realizing Rights Initiative, and the Global Company*. A resource guide to corporate human rights reporting, 2009. Retrieved from: http://www.globalreporting.org/NR/rdonlyres/4C5DB4C6–5084–4A84-BE51–0D134B3B5A2E/3583/HR_ReportFINAL_Resource_Guide.pdf (Accessed on 8 July 2019).

134. Hair, J. F., Black, W. C., Rabin, B. J., Anderson, R. E., & Tatham, R. L., (2006). *Multivariate Data*, 40–60.

135. Hamilton, V. L., (1978). Obedience and responsibility: A jury simulation. *Journal of Personality and Social Psychology, 36*(2), 126–146.

136. Hammond, A., William, J. K., Julia, T., et al., (2007). *The Next 4 Billion: Market Size and Business Strategy at the Base of the Pyramid*. World resources institute and international finance corporations, World Bank Group. Retrieved from: http://www.wri.org/publication/the-next–4-billion (Accessed on 8 July 2019).

137. Handfield, R., & Straube, F., & Pfohl, H. C., & Wieland, A., (2013). *Trends and Strategies in Logistics and Supply Chain Management,* 65–80.

138. Hansen, M. C., et al., (2013). High-resolution global maps of 21st-century forest cover change. *Science, 342*(6160), 850–853.

139. Harrison, J. S., Bosse, D. A., & Phillips, R. A., (2010). Managing for stakeholders. *Stakeholder*, 49–70.

140. Helliwell, J. F., Layard, R., & Sachs, J., (2015). *World Happiness Report 2015*. New York: Sustainable development solutions network.

141. Herman Miller, Inc., (2009). *Our Journey Toward a Better World Around You: A Report From Herman Miller, Inc. 2009* (pp. 39–52). Zeeland, Michigan.

142. Herz, S., Antonio, L. V., & Jonathan, S., (2010). *Development Without Conflict: The Business Case for Community Consent*. Retrieved from: http://www.wri.org/publication/developmentwithout-conflict (Accessed on 8 July 2019).

143. Hillman, A. J., & Keim, G. D., (2001) Shareholder value, stakeholder management, and social issues: What's the bottom line? *Strategic Management Journal, 22*, 125–139.

144. Hitt, M. A., Ireland, R. D., & Hoskisson, R. E., (2003). *Strategic Management: Competitiveness and Globalization*. Thomson South-Western, Mason, Ohio, 108–120.

145. Holder-Webb, L., Cohen, J. R., Nath, L., & Wood, D., (2009). The supply of corporate social responsibility disclosures among U.S. firms. *Journal of Business Ethics, 84*, 497–527.

146. Holdsworth, R., (2003). Practical applications approach to design, development, and implementation of an integrated management system. *Journal of Hazardous Materials, 104*(1), 193–205.

147. Hora, M., & Klassen, R., (2013). Learning from others' misfortune: Factors influencing knowledge acquisition to reduce operational risk. *Journal of Operations Management, 31*. 52–61.

148. Hora, M., Bapuji, H., & Roth, A. V., (2011). *Safety Hazard and Time to*

Recall: The Role of Recall, 47–90.

149. Huff, A. S., Floyd, S. W., Sherman, H. D., & Terjesen, S., (2009). *Strategic Management: Logic and Action*. John Wiley and Sons, Hoboken: NJ, 10–45.

150. IBM, (2008). "*A Measured Approach to Going Green: IBM 'Green Sigma' Consulting Offering to Help Clients Reduce Energy and Water Usage*." IBM.com. Retrieved at: http://www–935.ibm.com/ services/ us/gbs/bus/pdf/gbe03019-usen–02.pdf (Accessed on 8 July 2019).

151. IMF, (2015). *World Economic Outlook Database* (pp. 48–59). Washington, D.C.: International Monetary Fund.

152. Immelt, J., (2009). "*Chairman's Presentation to the General Electric Shareholders*." General electric annual shareholders meeting. Retrieved from: http://www.ge.com/pdf/investors/events/04222009/ ge_earnings_transcript_04222009_english.pdf (Accessed on 8 July 2019).

153. Immelt, J., (2009). "*Renewing American Leadership*." Distinguished leader series, black and gold forum. United States Military Academy at West Point, New York. Retrieved from: http://files.gereports.com/ wp-content/uploads/2009/12/90304–2-JRI-Speech-Reprint1–557. qxd_8.5x11.pdf (Accessed on 8 July 2019).

154. IPCC, (2007). "13.3.3.3– implications of regime stringency: Linking goals, participation and timing–Box 13.7," In: Metz, B., Davidson, O. R., Bosch, P. R., Dave, B. R., & Meyer, L. A., (eds.), *Contribution of Working Group III: Mitigation of Climate Change [Core Writing Team]*. Cambridge University Press, Cambridge, United Kingdom and New York, NY, USA. Retrieved from: http://www.ipcc.ch/ publications_ and _data/ar4/wg3/en/ch13-ens13–3–3–3.html (Accessed on 8 July 2019).

155. ITU, (2015). *World Telecommunication/ICT Indicators Database*. Geneva: International Telecommunication Union.

156. Iveco, (2009). "*Iveco-FedEx Express: First Results of the Daily Hybrid Evaluation*." Iveco.com. http://www.iveco.com/enus/PressRoom/ PressRelease/Pages/FedEx-apr09.aspx (Accessed on 8 July 2019).

157. Jamali, D., (2008). A stakeholder approach to corporate social responsibility: A fresh perspective into theory and practice. *Journal of Business Ethics, 82*(1), 213–231.

158. Jan, O. A., & Mattias, L., (2001). Ecologically unsustainable trade.

Ecological Economics, 37(1), 113–122.

159. Japan Environmental Management Association for Industry, (2009). *Carbon Footprint of Products*. Jemai.or.jp. Retrieved from: http://www.jemai.or.jp/english/carbonfootprint.cfm (Accessed on 8 July 2019).
160. Jayaraman, V., Linton, J., & Klassen, R., (2005). Sustainable supply chains: An introduction. *Journal of Operations Management, 2007*, 40–160.
161. Jennings, M. M., (2004). The role of purchasing in corporate social responsibility: A structural equation analysis. *Journal of Business Logistics, 25*(1), 145–186.
162. Jiang, B., (2009). Implementing supplier codes of conduct in global supply chains: Process explanations from theoretic and empirical perspectives. *Journal of Business*, 65–90.
163. Jonker, J., & Karapetrovic, S., (2004). Systems thinking for integration of management systems. *Business Process Management Journal, 10*(06), 608–615.
164. Jørgensen, T. H., (2008). Towards more sustainable management systems: Through life cycle management and integration. *Journal of Cleaner Production, 16*(10), 1071–1080.
165. Jørgensen, T. H., Remmen, A., & Mellado, M. D., (2006). Integrated management systems three different levels of integration. *Journal of Cleaner Production, 14*(8), 713–722.
166. Kalamas, M., Cleveland, M., & Laroche, M., (2014). *Pro-Environmental Behaviors for Thee*, 38–61.
167. Karapetrovic, S., (2002). Strategies for the integration of management systems and standards. *The TQM Magazine, 14*(1), 61–67.
168. Karapetrovic, S., (2003). Musings on integrated management systems. *Measuring Business Excellence, 7*(1), 4–13.
169. Karapetrovic, S., & Jonker, J., (2003). Integration of standardized management systems: Searching for a recipe and ingredients. *Total Quality Management, 14*(4), 451–459.
170. Karapetrovic, S., & Willborn, W., (2001). Audit and self-assessment in quality management: Comparison and compatibility. *Managerial Auditing Journal, 16*(6), 366–377.
171. Kats & Gregory, (2003). *Green Building Costs and Financial Benefits*.

Published in USA Massachusetts technology collaborative. Retrieved from: http://www.noharm.org/lib/downloads/building/ Building_Green_Costs_Benefits_Sum.pdf (Accessed on 8 July 2019).

172. Keeble, J., Topiol, S., & Berkeley, S., (2003). Using indicators to measure sustainability performance at a corporate and project level. *Journal of Business Ethics, 44*, 149–158.

173. Keegan & Paul, (2009). "*Zipcar: The Best New Idea in Business.*" Fortune. Retrieved from: http://money.cnn.com/2009/08/26/ news/companies/zipcar_car_rentals.fortune/ (Accessed on 8 July 2019).

174. Key, S., & Popkin, S. J., (1998). Integrating ethics into the strategic management process: Doing well by doing good. *Management Decision, 36*(5), 331–338.

175. Kim, E. H., & Lyon, T. P., (2011). "Strategic environmental disclosure: Evidence from the DOE's voluntary greenhouse gas registry." *Journal of Environmental Economics and Management, Elsevier, 61*(3), 311–326.

176. Klassen & Vereecke, (2012), Social issues in supply chains: Capabilities link responsibility, risk (opportunity), and performance. *International Journal of Production Economics, 140*(1), 103–115.

177. Klein, J., & Dawar, N., (2004) Corporate social responsibility and consumers' attributions and brand evaluations in a product-harm crisis. *International Journal of Research in Marketing, 21*, 203–217.

178. Kleine, A., & Hauff, M. V., (2009). Sustainability-driven implementation of corporate social responsibility: Application of the integrative sustainability triangle. *Journal of Business Ethics, 85*(3), 517–533.

179. Kouabenan, D. R., Medina, M., Gilibert, D., & Bouzon, F., (2001). Hierarchical position, gender, accident severity, and causal attribution. *Journal of Applied Social Psychology*, 120–140.

180. Kovács, G., (2008). Corporate environmental responsibility in the supply chain. *Journal of Cleaner Production, 16*(15), 1571–1578.

181. Kroll, C., (2015). Sustainable development goals: Are the rich countries ready? *Gutersloh: Bertelsmann Foundation*, 37–90.

182. Kurucz, E., Colbert, B., & Wheeler, D., (2008). The business case for corporate social responsibility. In: Crane, A., McWilliams, A., Matten, D., Moon, J., & Siegel, D., (eds.), *The Oxford Handbook of Corporate Social Responsibility* (pp. 83–112). Oxford: Oxford University Press. (Google Scholar).

183. Lee, M. D. P., (2008). A review of the theories of corporate social responsibility: Its evolutionary path and the road ahead. *International Journal of Management Reviews, 10*(1), 53–73 (View Article Google Scholar).

184. Lee, S. Y., & Klassen, R. D., (2008). *Drivers and Enablers That Foster Environmental Management Capabilities in Small- and Medium-Sized Suppliers in Supply Chains*, 25–67.

185. Levitt, T., (1958). *The Dangers of Social Responsibility* (pp. 41–50). Harvard Business Review.

186. Li, M., Choi, T. Y., Rabinovich, E., & Crawford, A., (2013). *Self-Service Operations at Retail*, 12–32.

187. Lohman, C., Fortuin, L., & Wouters, M., (2004). Designing a performance measurement system: A case study. *European Journal of Operational Research, 156*(2), 267–286.

188. Lydenberg, S., & Katie, G., (2008). *Innovations in Social and Environmental Disclosure Outside the United States*. Domini social investments and social investment forum. Retrieved from: http://www.domini.com/common/pdf/Innovations_in_ Disclosure.pdf (Accessed on 8 July 2019).

189. Macey, J. R., (1997). "Institutional investors and corporate monitoring: A demand-side perspective: Introduction." *Managerial and Decision Economics (1986–1998), 18*(7/8), p. 601.

190. MacKenzie, S. B., & Podsakoff, P. M., (2012). *Common Method Bias in Marketing: Causes*, 56–90.

191. Madden, J., (2012). "Operational activism hits the boardroom: One of the most important trends in governance is the increasing activity of shareholders pressuring boards not on classic governance subjects but on the actual strategic direction or management of the business of the corporation." *Directors and Boards, 37*(1), p. 37.

192. Makni, R., Francoeur, C., & Bellavance, F., (2009). Causality between corporate social performance and financial performance: Evidence from Canadian firms. *Journal of Business Ethics, 89*, 409–422.

193. Malik, O., (2013). *Global Database of Toil Wastewater Treatment* (pp. 56–71). New Haven, CT: Yale Center for Environmental Law and Policy.

194. Maon, F., Lindgreen, A., & Swaen, V., (2009). Designing and implementing corporate social responsibility: An integrative framework

grounded in theory and practice. *Journal of Business Ethics, 87*(1), 71–89.

195. Maon, F., Lindgreen, A., & Swaen, V., (2010). Organizational stages and cultural phases: A critical review and a consolidative model of corporate social responsibility development. *International Journal of Management Review, 12*(1), 20–38.
196. Margolis, J., & Walsh, J., (2003). Misery loves companies: Rethinking social initiatives by business. *Administrative Science Quarterly, 48*(2), 268–305.
197. Martin & Andrew, (2007). "*In Eco-Friendly Factory, Low-Guilt Potato Chips*." NYTimes.com. Retrieved from: http://www.nytimes.com/2007/11/15/business/15plant.html (Accessed on 8 July 2019).
198. Mathers & Jason, (2009). "*Corporate Fleets Continue to Innovate for Climate Reductions*." Greenbiz.com. Retrieved from: http://www.greenbiz.com/blog/2009/12/07/corporate-fleets-continueinnovate-emission-reductions (Accessed on 8 July 2019).
199. Matos, S., & Hall, J., (2007). *Integrating Sustainable Development in the Supply Chain*, 43–90.
200. McDonald, M., Mors, T. A., & Phillips, A., (2003). Management system integration: Can it be done? *Quality Progress, 36*(10), 67–74.
201. McDonough, W., & Michael, B., (2002). *Cradle to Cradle* (pp. 1–15). North Point Press.
202. McGiboney, M., (2009). "*Twitter's Tweet Smell of Success*." Nielsen wire. Retrieved from: http://blog.nielsen.com/nielsenwire/online_mobile/twitters-tweet-smell-of-success/ (Accessed on 8 July 2019).
203. McGraw Hill Construction and Siemens, (2009). *Greening of Corporate America: The Pathway to Sustainability*. Retrieved from: http://construction.com/market_research/FreeReport/GreeningCorpAmerica/GreeningCorporateAmerica.asp (Accessed on 8 July 2019).
204. McKinsey and Company, (2008). *McKinsey Global Survey Results: Valuing Corporate Social Responsibility*. McKinsey Quarterly. Retrieved from: commdev.org/files/2393_file_McKQ_Valuing_Corporate_Social_ Responsibility.pdf (Accessed on 8 July 2019).
205. McKinsey and Company, (2009). *Unlocking Energy Efficiency in the US Economy*. Retrieved from: http://www.mckinsey.com/ clientservice/electricpowernaturalgas/US_energy_efficiency/ (Accessed on 8 July

2019).

206. McKinsey Company, (2009). *The Global Corporate Water Footprint: Risks, Opportunities, and Management Options*. Retrieved from: http://www.mckinsey.com/clientservice/ccsi/pdf/Report_Large_Water_Users.pdf (Accessed on 8 July 2019).

207. Melnyk, S. A., Sroufe, R. P., & Calantone, R., (2003). Assessing the impact of environmental management systems on corporate and environmental performance. *Journal of Operations Management, 21*(3), 329–351.

208. Mena, C., Humphries, A., & Choi, T., (2013). Toward a theory of multi-tier supply chain management. *Journal of Supply Chain Management, 49*(2), 58–77.

209. Meyer & Tobias, A., (2009). *Energy Efficient Supply Chains*. McKinsey Quarterly. Retrieved from: http://www.mckinseyquarterly.com/Increasing_the_energy_efficiency_of_supply_chains_2414?pagenum=1#interactive (Accessed on 8 July 2019).

210. Mishra, R. K., & Jhunjhunwala, (2013). *SBooks24x7, I Diversity and the Effective Corporate Board* (1st edn.), Academic Press, US.

211. Mitchell, R. K., Agle, B. R., & Wood, D. J., (1997). Towards a theory of stakeholder identification and salience: Defining the principle of who and what really counts. *Academy of Management Review, 22*(4), 853–886.

212. Moneva, J. M., Archel, P., & Correa, C., (2006). GRI and the camouflaging of corporate unsustainability. *Accounting Forum, 30*(2), 121–137.

213. Monks, R. A. G., & Minow, N., (2012). *Corporate Governance* (5th edn., 12–45). John Wiley and Sons.

214. Muthén, L. K., & Muthén, B. O., (2010). Mplus. *Statistical Analysis with Latent Variables* (6th edn.). User's Guide, Muthén and Muthén, Los Angeles, CA.

215. Muthén, L. K., & Muthén, B. O., (2013). *Chi-Square Difference Testing Using the Santora–Bentler Scaled Chi-Square*. http://www.statmodel.com/chidiff.shtml (Accessed on 8 July 2019).

216. NALGEP and Smart Growth Leadership Institute, (2004). *Smart Growth is Smart Business*. Washington, DC.

217. Narasimhan, R., & Talluri, S., (2009). Perspectives on risk management in supply chains. *Journal of Operations Management, 27*(2), 114–118.

218. Natural Resources Defense Council, (2009). "*Green Business: Green Business Guides*." Retrieved from: http://www.nrdc.org/enterprise/greeningadvisor/ta-idling.asp (Accessed on 8 July 2019).

219. New, S., (2004). The ethical supply chain. In: New, S., & Westbrook, R., (eds.), *Understanding Supply Chains: Concepts, Critiques, and Futures* (pp. 253–280). Oxford University Press, Oxford, UK.

220. Nicolai, S., Hoy, C., Berliner, T., & Thomas A., (2015). *Projecting Progress: Reaching the SDGs by 2030* (35–62). London: Overseas Development Institute.

221. Nijhof, A., Bruijn, T. D., Fisscher, O. A. M., Jonker, J., Karssing, E., & Schoemaker, M., (2005). Learning to be responsible: Developing competencies organization-widen wide CSR. In: Jonker, J., & Cramer, J., (eds.), *Making a Difference: Dutch National Research* (45–78).

222. Noe, T. H., (2002). "Investor activism and financial market structure." *The Review of Financial Studies, 15*(1), pp. 289–318.

223. Nye, J. S., (1974). Multinational corporations in world politics. *Foreign Affairs, 53*(1), 153e175.

224. O'Rourke, A., (2003). "A new politics of engagement: Shareholder activism for corporate social responsibility." *Business Strategy and the Environment, 12*(4), p. 227.

225. OECD, (2004). "OECD principles of corporate governance." *Journal of Economic Literature, 42*(4), p. 1199.

226. OECD, (2016). OECD stat. *Paris: Organization for Economic Co-Operation and Development*, 24–28.

227. Organization for economic co-operation and development, (2009). "*Fact Sheet: Extended Producer Responsibility." Environ Directorate*. Retrieved from: http://www.oecd.org/document/53/0, 3343, en_2649_34395_37284725_1_1_1_1,00.htm in November 2009 (Accessed on 8 July 2019).

228. Orlitzky, M., Schmidt, F., & Rynes, S., (2003). Corporate social and financial performance: A meta-analysis. *Organization Studies, 24*(03), 403–441.

229. Oskarsson, K., & Malmborg, F. V., (2005). Integrated management systems as a corporate response to sustainable development. *Corporate Social Responsibility and Environmental Management, 12*(3), 121–128.

230. Parmigiani, A., Klassen, R. D., & Russo, M. V., (2011). Efficiency

meets accountability: Performance implications of supply chain configuration, control, and capabilities. *Journal of Operations Management, 29*(3), 212–223.

231. Pedersen, E. P., (2006). Making corporate social responsibility (CSR) operable: How companies translate stakeholder dialogue into practice. *Business and Society Review, 111*(2), 137–163.
232. Penn, S., Berland, A. L., & Burson, M., (2009). *Corporate Citizenship Study*. Retrieved from: http://www.burson-marsteller.com/Innovation_and_insights/blogs_and_podcasts/BM_Blog/Documents/Corporate Citizenship Executive Summary.pdf (Accessed on 8 July 2019).
233. Phares, E. J., & Wilson, K. G., (1972). Responsibility attribution: Role of outcome severity, situational ambiguity internal-external control. Journal of Personality, 40(3), 392–406.
234. Phillips, R., & Caldwell, C. B., (2005). Value chain responsibility: A farewell to arm's length. *Business and Society Review, 110*(4), 345–370.
235. Pinkston, T. S., & Carroll, A. B., (1996). A retrospective examination of CSR orientations: Have they changed? *Journal of Business Ethics, 15*(2), 199–206 (View Article Google Scholar).
236. Podsakoff, P. M., MacKenzie, S. B., Lee, J. Y., & Podsakoff, N. P., (2003). Common method biases in behavioral research: A critical review of the literature and recommended remedies. *Journal of Applied Psychology, 88*(5), 879–904.
237. Pohle, G., & Jeff, H., (2008). *Attaining Sustainable Growth Through Corporate Social Responsibility.* IBM Institute for Business Value. Retrieved from: http://www–935.ibm.com/services/us/gbs/bus/pdf/gbe03019-usen–02.pdf (Accessed on 8 July 2019).
238. Price Water House-Coopers, LLP and Carbon Disclosure Project, (2009). *Supply chain report 2009*. CDProject.net. Retrieved from: https://www.cdproject.net/CDPResults/65_329_201_CDP-Supply-ChainReport_2009.pdf (Accessed on 8 July 2019).
239. Program on corporate social responsibilities, (2000). *Ministry of Economic Affairs* (p. 57e84). The Hague.
240. Rehbein, K., Waddock, S., & Graves, S. B., (2004). "Understanding shareholder activism: Which corporations are targeted?" *Business and Society, 43*(3), p. 239.
241. Renewable Energy World.com., (2009). "*Installed US Wind*

Energy Capacity Grows by Record 8,300 MW." Washington D. C. Retrieved from: http:/www.renewableenergyworld.com/rea/news/article/2009/01/installed-us-wind-energy-capacity-grows-by-record–8300-mw–54619 (Accessed on 8 July 2019).

242. Reuter, C., Foerstl, K., Hartmann, E., & Blome, C., (2010). Sustainable global supplier management: The role of dynamic capabilities in achieving competitive advantage. *Journal of Supply Chain Management: A Global Review of Purchasing and Supply, 46*(2), 45–63.
243. Rickels, W., Quaas, M. F., & Visbeck, M., (2014). How healthy is the human-ocean system? *Environmental Research Letters*, 9. doi: 10.1088/1748–9326/9/4/044013.
244. Risk Metrics Group, (2008). *Governance White Paper: One in Five Large Firms Set Labor Supplier Standards*. Retrieved from: http://www.riskmetrics.com/node/135625 (Accessed on 8 July 2019).
245. Robeco-Booz and Company, (2008). *Responsible Investing: A Paradigm Shift–From Niche to Mainstream*. Retrieved from: http://investmentadvisor.com/Issues/2009/September%202009/PublishingImages/Whitepaper_ResponsibleInvesting.pdf (Accessed on 8 July 2019).
246. Rocha, M., Searcy, C., & Karapetrovic, S., (2007). Integrating sustainable development into existing management systems. *Total Quality Management and Business Excellence, 18*(1/2), 83–92.
247. Rodrigues, A. S., Brooks, T. M., Butchart, S. H., Chanson, J., Cox, N., Hoffmann, M., & Stuart, S. N., (2014). Spatially explicit trends in the global conservation status of vertebrates. *Plus One, 9*(11), e113934.
248. Roehm, M. L., & Tybout, A. M., (2006). When will a brand scandal spill over, and how should competitors respond? *Journal of Marketing Research, 43*(3), 366–373.
249. Roth, A. V., Tsay, A. A., Pullman, M. E., & Gray, J. V., (2008). Unraveling the food supply chain: Strategic insights from China and the 2007 recalls. *Journal of Supply Chain Management, 44*(1), 22–39.
250. Rowley, T. I., & Moldoveanu, M., (2003). When will stakeholder groups act? An interest-based model of stakeholder group mobilization. *Academy of Management Review, 28*(2), 204–219.
251. Ruggie and John, (2006). *Human Rights Policies and Management Practices of Fortune Global 500 Firms: Results of a Survey*. Retrieved

from: http://www.reports-and-materials.org/Ruggie-surveyFortune-Global 500.pdf.

252. Salomone, R., (2008). Integrated management systems: Experiences in Italian organizations. *Journal of Cleaner Production, 16*(16), 1786–1806.
253. Salzmann, O., Iones, A., & Steger, U., (2005). The business case for corporate sustainability: Literature review and research options. *European Management Journal, 23*(1), 27–36.
254. San Diego Association of Governments (SANDAG), (2006). "*Smart Growth Definition, Principles, and Designations*." Retrieved from: http://www.lisc.org/san_diego/assets/asset_upload_file873_6802.pdf (Accessed on 8 July 2019).
255. Sarkis, J., (2004). Relationships between operational practices and performance among early adopters of green supply chain management practices in Chinese manufacturing enterprises. *Journal of Operations Management, 22*(3), 25–65.
256. Sarzynski, A., Marilyn, B., & Frank S., (2008). *Shrinking the Carbon Footprint of Metropolitan America Brookings Institution*. Retrieved from: http://www.brookings.edu/ reports/2008/05_carbon_footprint_sarzynski.aspx (Accessed on 8 July 2019).
257. Schwartz, M., & Carroll, A. B., (2003). Corporate social responsibility three-domain approach. *Business Ethics Quarterly, 13*(4), 503–530 (View Article Google Scholar).
258. Schwartz, S. H., & Bilsky, W., (1990). Toward a theory of the universal content and structure of values: Extensions and cross-cultural replications. *Journal of Personality and Social Psychology, 58*(1990), 550–562.
259. Scott, L., (2009). "*Chandler Company to Help McDonald's Cut Energy Use, Cost*." The Arizona Republic. Retrieved from: http://www.azcentral.com/community/chandler/articles/2008/12/31/20081231cr-engenuity0102.html?&wired (Accessed on 8 July 2019).
260. SDSN, (2015). *Indicators and a Monitoring Framework for Sustainable Development Goals: Launching a Data Revolution for the SDGs*. New York and Paris: Sustainable Development Solutions Network.
261. Searcy, C., McCartney, D., & Karapetrovic, S., (2007). Sustainable development indicators for the transmission system of an electric utility. *Corporate Social Responsibility and Environmental Management,*

14(3), 135–151.

262. Seawright, K. K., & Sampson, S. E., (2007). A video method for empirically studying wait-perception bias. *Journal of Operations Management, 25*(5), 1055–1066.
263. Sen, S., & Battacharya, C., (2001). Does doing good always lead to doing better? Consumer reactions to corporate social responsibility. *Journal of Marketing Research, 38*(2), 225–243.
264. Seuring, S., & Mueller, M., (2008). From a literature review to a conceptual framework for sustainable supply chain management. *Journal of Cleaner Production, 16,* 1699–1710.
265. Shaver, K., (1985). *The Attribution of Blame: Causality, Responsibility, and Blameworthiness*. Springer, New York, NY, 54–78.
266. Skarmeas, D., & Leonidou, C. N., (2013). When consumers doubt, watch out! The role of CSR skepticism. *Journal of Business Research, 66*(10), 1831–1838.
267. Slater & Allison, (2008). *KPMG International Survey of Corporate Responsibility Reporting 2008*. KPMG.com. Retrieved from: http://us.kpmg.com/RutUS_prod/Documents/8/Corporate_Sustainability_Report_US_Final.pdf (Accessed on 8 July 2019).
268. Smith, N. C., (2003). Corporate social responsibility: whether or how? *California Management Review, 45*(4), 52–76.
269. Smith, W. J., Wokutch, R. E., Harrington, K. V., & Dennis, B. S., (2001). An examination of the influence of diversity and stakeholder role on corporate social orientation. *Business and Society, 40*(3), 266–294 (View Article Google Scholar).
270. Social Accountability International, (2009). "*SAI Training Programs*." Retrieved from: http://www.sa-intl.org/index.cfm?fuseaction=Page.vie wPage&pageId=553&parentID=473&nodeID=1 (Accessed on 8 July 2019).
271. Social Investment Forum, (2007). *Report on Socially Responsible Investing Trends in the United States*. Retrieved from: http://www.socialinvest.org/pdf/SRI_Trends_ExecSummary_2007.pdf on 20 January 2010 (Accessed on 8 July 2019).
272. Sodexo, (2008). *Sustainable Food Procurement*. Dublin. Retrieved from: http://uk.sodexo.com/uken/Images/SustainableProcurementbrochure_tcm15–160012.pdf (Accessed on 8 July 2019).
273. Sodhi, M. S., Son, B. G., & Tang, C. S., (2012). Researchers'

perspectives on supply chain risk management. *Production and Operations Management, 21*(1), 1–13.

274. Solomon, J., (2010). *Corporate Governance and Accountability*. Wiley, West Sussex, 68–108.
275. Spence, L., (2016). Small business social responsibility: Expanding core CSR theory. *Business and Society, 55*(1), 23–55 (View Article Google Scholar).
276. Sperling and Daniel, (2009). "*Biofuels and the LCFS*." Retrieved from: http://biomass.ucdavis.edu/materials/forums%20and%20workshops/f2009/0.2_Dan%20Sperling.pdf (Accessed on 8 July 2019).
277. Staples, (2006). "*Energy and Climate*." Staples.com. Retrieved from: http://www.staples.com/sbd/content/about/soul/energyclimate.html (Accessed on 8 July 2019).
278. Steiger, J. H., (2007). Understanding the limitations of global fit assessment in structural equation modeling. *Personality and Individual Differences, 42*(5), 893–898.
279. Steurer, R., Langer, M. E., Konrad, A., & Martinuzzi, A., (2005). Corporations, stakeholders and sustainable development I: A theoretical business-society relations. *Journal of Business Ethics, 61*(3), 263–281.
280. Stiglitz, J. E., Jonathan, M. O., & Peter, R. O., (2006). *The Impact of Asbestos Liabilities on Workers in Bankrupt Firms*. Sebago Associates, commissioned by the American Insurance Association. Washington, DC, 2002.
281. Struthers, C. W., Eaton, J., Ratajczak, A., & Perunovic, M., (2004). *Social Conduct Toward Sumer Attributions of Blame: Defense Attributions in Product-Harm Crises in*, 1–17.
282. Suncor Energy, (2006). "*Excellence in Reporting*." Retrieved from: http://www.suncor.com/en/responsible/1879.aspx (Accessed on 8 July 2019).
283. Suncor Energy, (2008). *Climate Change, We're Listening, We're Responding: 2008 Progress Report on Climate Change*. Retrieved from: http://www.suncor.com/pdf/climate-change-report–2008.pdf (Accessed on 8 July 2019).
284. Surroca, J., Tribó, J. A., & Waddock, S., (2010). Corporate responsibility and financial performance: The role of intangible resources. *Strategic Management Journal, 31*(5). 17–46.
285. Swanson, D. L., (1995). Addressing a theoretical problem by reorienting

the corporate social performance model. *Academy of Management Review, 20*, 43–64 (Google Scholar).

286. Swiss Re., (2008). "*Swiss Re Staff Actively Contribute to Reducing CO2 Emissions*." SwissRe.com. Retrieved from: http://www.swissre.com/pws/media%20centre/news/news%20releases%202008/coyou2_release_january_2008_1_year.htm (Accessed on 8 July 2019).
287. Tennen, H., & Affleck, G. (1990). Blaming others for threatening events. *Psychological Bulletin, 108*(2), 209–232.
288. The Hackett Group, (2006). "*The Hackett Group: Supplier Diversity Does Not Drive Increased Costs*." Retrieved from: http://www.thehackettgroup.com/about/alerts/alerts_2006/alert_08172006.jsp (Accessed on 8 July 2019).
289. The International Council on Clean Transportation, (2006). *Air Pollution and Greenhouse Gas Emissions from Ocean-Going Ships: Impacts, Mitigation, Options and Opportunities for Managing Growth*. Retrieved from: http://www.theicct.org/ (Accessed on 8 July 2019).
290. The Triple Bottom Line, (2006). *Does It All Add Up?* (pp. 1–16). Earthscan, London, UK.
291. The United States Conference of Mayors, (2009). *Taking Local Action: Mayors and Climate Protection Best Practices*. Retrieved from: http://www.usmayors.org/pressreleases/uploads/ClimateBestPractices061209.pdf (Accessed on 8 July 2019).
292. Transparency International, (2015). *Corruption Perception Index 2014 Results*. Berlin: Transparency International.
293. U.S. Environmental Protection Agency (EPA), (2006). Executive summary: Inventory of U.S. *Greenhouse Gas Emissions and Sinks: 1990–2007*. http://www.epa.gov/climatechange/emissions/downloads09/ExecutiveSummary.pdf (Accessed on 8 July 2019).
294. U.S. Environmental Protection Agency, (2006). "*Essential Principles for Reform of Chemicals Management Legislation*." EPA.gov. Retrieved at: http://www.epa.gov/oppt/ existingchemicals/pubs/principles.html (Accessed on 8 July 2019).
295. U.S. Environmental Protection Agency, (2006). "Green Suppliers Network." Retrieved from: http://www.epa.gov/oppt/ar/2007–2009/workig/gsn.htm (Accessed on 8 July 2019).
296. U.S. Environmental Protection Agency, (2006). *Design for the Environment Partnership Highlights*. Retrieved from: http://www.

epa.gov/dfe/pubs/index.htm (Accessed on 8 July 2019). 130. Kanter, Rosa Beth Moss. "SUPERCORP: How vanguard companies create innovation, profits, growth, & social good, 2010."

297. U.S. Environmental Protection Agency, (2006).*Green Chemistry*. Summary of 2007 award entries and recipients. Presidential green chemistry awards. Washington, DC, 2007. Retrieved from: http://www.epa.gov/opptintr/greenchemistry/pubs/pgcc/winners/dgca07.html (Accessed on 8 July 2019).
298. U.S. Green Building Council, (2006). "*Genzyme Project Overview*." Retrieved from: http://leedcasestudies.usgbc.org/orview.cfm?ProjectID=274 (Accessed on 8 July 2019).
299. UN Environment Program Sustainable Energy Finance Initiative (SEFI) and New Energy Finance, (2006). *Global Trends in Sustainable Energy Investment 2009*. Retrieved from: http://sefi.unep.org/english/globaltrends2009.html (Accessed on 8 July 2019).
300. UN Webcast, (2006). "*Summary Report to the Human Rights Council of the UN by the Special Employee Secretary-General on Transnational Corporations and Human Rights*." A/HRC/11/13. Retrieved from: http://www.un.org/webcast/unhrc/ archive.asp?go=011 (Accessed on 8 July 2019).
301. UN-Water, (2006). *Coping With Water Scarcity: Challenge of the Twenty-First Century*. Retrieved from: http://www.fao.org/nr/water/docs/escarcity.pdf (Accessed on 8 July 2019).
302. UN, (2015). *Transforming Our World: The 2030 Agenda for Sustainable Development*. New York: United Nations.
303. UNDP, (2015). *Human Development Statistical Tables*. Human development report 2015. New York: United Nations Development Programme.
304. UNEP, (2009). *Finance Initiative Asset Management Working Group*. Fiduciary responsibility: Legal and practical aspects of integrating environmental, social and governance issues into institutional investment. Retrieved from: http://www.unepfi.org/fileadmin/documents/fiduciaryII.pdf (Accessed on 8 July 2019).
305. UNEP, (2016). *UNEP Live Natural Resources: Resource Efficient Indicators*. Nairobi: United Nations Environment Program me.
306. UNESCO, (2015). *UIS Database*. Paris: United Nations Educational, Scientific and Cultural Organization.

307. Unilever, (2009). "*Environmental Responsibility: Water.*" Unilever. com. Retrieved from: http://www.unilever.com/sustainability/environment/water/?WT.LHNAV=Water (Accessed on 8 July 2019).

308. Unilever, (2010). "*India: Lifebuoy Promotes Hand Washing with Soap to Improve Health.*" Retrieved from: http://www. unilever. com/sustainability/casestudies/health-nutrition-hygiene/indialifebuoypromoteshandwashingwithsoaptoimprovehealth.aspx (Accessed on 8 July 2019).

309. United Nations Global Compact, (2017). "*CEO Water Mandate.*" The foundation for the global compact. Retrieved from: http://www. unglobalcompact.org/Issues/Environment/CEO_Water_Mandate/ (Accessed on 8 July 2019).

310. United Nations, Global Compact, (2017)."*CEO Statement: 60th Anniversary of the Universal Declaration for Human Rights.*" UNGlobalCompact.org. 20 November 2008. Retrieved from: http://www.unglobalcompact.org/Issues/human_rights/CEO_Statemen_List.html (Accessed on 8 July The 2019).

311. United States, (2004). *Department of Energy, Industrial Technologies Program*. Best practices: Corporate energy management case study. WashingtoThe n, DC.

312. United States, National Bureau of Economic Research, (2005). *Morck, R. K.: A History of Corporate Governance Around the World*. University of Chicago The Press.

313. United States, National Bureau of Economic Research, (2007). *Transparency and Corporate Governance*. Massachusetts: NBER (Working Paper 1The 2875).

314. United States, National Bureau of Economic Research, (2008). *The Role of Boards of Directors in Corporate Governance: A Conceptual Framework and Survey*. Massachusetts: NBER (Working Paper 1The 4486).

315. United States, National Bureau of Economic Research, (2009). *The State of Corporate Governance Research*. Massachusetts: NBER (Working Paper 15537).

316. UPS, (2008). "*Letter from the Chairman.*" 2008 UPS corporate sustainability report. 2009. 20 January 2010 http://www.sustainability. ups.com/community/Static%20Files/ sustainability/2008_CSR_PDF_Report.pdf (Accessed on 8 July 2019).

317. UPS, (2008). *UPS Corporate Sustainability Report*. Retrieved from: http://www.sustainability.ups.com/community/Static%20Files/sustainability/2008_CSR_PDF_Report.pdf (Accessed on 8 July 2019).

318. Urda, J., & Loch, C. H., (2013). Social preferences and emotions as regulators of behavior utility functions, and competitive advantage. *Strategic Management Journal, 31*, 56–74.

319. Vachon, S., & Klassen, R. D., (2006). Green project partnership in the supply chain: The case of the package printing industry. *Journal of Cleaner Production 14*(6/7), 49–120.

320. Vachon, S., & Klassen, R. D., (2008). Environmental management and manufacturing performance: The role of collaboration in the supply chain. *International Journal of Production Economics, 111*, 299–315.

321. Van Burg, E., & Van Oorschot, K. E., (2013). Cooperating to commercialize technology: A dynamic model of fairness perceptions, experience and cooperation. *Production and Operations Management, 22*, 114–128.

322. Van der Heijden, A., Driessen, P. P. J., & Cramer, J. M., (2010). Making sense of corporate social responsibility: Exploring organizational processes and strategies. *Journal of Cleaner Production, 18*(18), 1787–1796.

323. Varmazis, M., (2008). "*Green Procurement Goes into the Black.*" Purchasing.com. http://www.purchasing.com/article/220103-_Green_procurement_goes_into_the_black.php (Accessed on 8 July 2019).

324. Visser, W., (2006). Revisiting Carroll's CSR pyramid: An African perspective. In: Huniche, M., & Rahbek, E. P., (eds.), *Corporate Citizenship in Developing Countries—New Partnership Perspectives* (pp. 29–56). Copenhagen: Copenhagen Business School Press. Google Scholar.

325. Visser, W., (2011). *The Age of Responsibility: CSR 2.0 and the New DNA of Business*. West Sussex: John Wiley and Sons. Google Scholar.

326. Walsh, K., (2008). *KLD's Global Climate 100 Index Marks Third Anniversary*. Risk metrics group ESG insight. Retrieved from: http://blog.kld.com/uncategorized/kld%E2%80%99s-globalclimate–100-index-marks-third-anniversary/ (Accessed on 8 July 2019).

327. Wartick, S. L., & Cochran, P. L., (1985). The evolution of the corporate social performance model. *Academy of Management Review, 10*, 758–769. Google Scholar.

328. WBCSD, (2000). *Corporate Social Responsibility: Making Good Business Sense*. World Business Council for Sustainable Development, Geneva.

329. Weber, M., (2008). The business case for corporate social responsible company-level measurement approach for CSR. *European Management Journal, 26*(4), 247–261.

330. WEF, (2015). *The Global Competitiveness Report 2015–2016*. Colony, Switzerland: World Economic Forum.

331. Weiner, B., (1986). An attributional theory of achievement motivation and emotion. *Psychological Review, 92*(4), 548–573.

332. Weiner, B., (1995). *Judgments of Responsibility: A Foundation for a Theory of Social Conduct* (pp. 26–90). New York: Guilford.

333. Welford, R., & Gouldson, A., (1993). *Environmental Management and Business Strategy*. Pitman, Boston: MA.

334. Whitehead, J., (2009). "*CRO/CSO Data Trend*." E-mail to Natasha Scot Nicki.

335. WHO, (2016a). *Global Health Observatory*. Geneva: World Health Organization.

336. WHO, (2016b). *Global Database on Body Mass Index*. Geneva: World Health Organization.

337. Williams, P., (2009). "*Written Testimony of IBM to the US Senate Committee on Energy and Natural Resources, in Support of a Hearing on the Energy-Water Nexus*." Email to Brooke Barton.

338. Wood, D. J., (1991). Corporate social responsibility revisited. *Academy of Management Review, 16*, 691–718. Google Scholar.

339. Wood, D., & Jones, R., (1996). Research in corporate social performance: What have we learned. In: Dwight, B., & Dennis, Y., (eds.), *Corporate Philanthropy at the Crossroads*. Bloomington: Indiana University Press. Google Scholar.

340. World Bank, (2015). *Country and Lending Groups*. Washington D. C. World Bank.

341. World Bank, (2016). *World Development Indicators*. Washington, D. C. World Bank.

342. World Business Council for Sustainable Development (WBSCD), (2009). *Transforming the Market: Energy Efficiency in Buildings*. Retrieved from: http://www.wbcsd.org/Plugins/DocSearch/details.as

p?DocTypeId=25&ObjectId=MzQyMDQ (Accessed on 8 July 2019).

343. World Economic Forum, (2009). *Supply Chain Decarbonization: The Role of Logistics and Transport in Reducing Supply Chain Carbon Emissions*. Retrieved from: http://www.weforum.org/pdf/ip/SupplyChainDecarbonization.pdf (Accessed on 8 July 2019).

344. Wright, T., (2000). IMS three into one: The advantages of a single integrated quality, health and safety, and environmental management system. *The Quality Assurance Journal, 4*(3), 137–142.

345. Wu, Z., & Pagell, M., (2011). *Balancing Priorities: Decision-Making in Sustainable Supply*, 120–160.

346. Xerox, (2006). "*Xerox's Emulsion Aggregation Toner–An Environmentally Friendly Technology*." Retrieved from: http://www.fujixerox.com.au/environment/design_environment.jsp (Accessed on 8 July 2019).

347. Yoon, Y., Gürhan-Canli, Z., & Schwarz, N., (2006). The effect of corporate social responsibility (CSR) activities on companies with bad reputations. *Journal of Consumer York, NY*, 140–220.

348. Young, T., (2009). "*US Energy Firms Commit to Electric Vehicle Fleets*." BusinessGreen.com. Retrieved from: http://www.businessgreen.com/business-green/news/2250177/energy-firmscommit-electric (Accessed on 8 July 2019).

349. Zhu, Q., & Sarkis, J., (2004) Relationships between operational practices and performance among early adopters of green supply chain management practices in Chinese manufacturing enterprises. *Journal of Operations Management, 22*, 265–289.

350. Ziek, P., (2009). Making sense of CSR communication. *Corporate Social Responsibility and Environmental Management, 16*(3), 137–145.

351. Zutshi, A., & Sohal, A. S., (2005). Integrated management system: The experiences of three Australian organizations. *Journal of Manufacturing Technology Management, 16*(2), 211–232.

352. Zwillinger, D., & Kokoska, S., (2000). *CRC Standard Probability and Statistics Tables and Formulae* (pp. 38–92). Chapman and Hall: New York. Section 14.7.

353. Zyglidopoulos, S. C., (2001). The impact of accidents on firms' reputation for social performance. *Business and Society, 40*(4), 416–441.

354.
355.
356. **Website and Blogs**
357. Are You Ready? Here Are The Top 10 Skills For The Future, Reuven Gorsht, SAP, (2014). http://www.forbes.com/sites/sap/2014/05/12/are-you-ready-here-are-the-top–10-skills-for-the-future/ (Accessed on 8 July 2019).
358. Arnim, W., Lauren, W., & Charles, L. R., (2010). *Key Competencies in Sustainability: A Reference Framework for Academic Program Development*. http://link.springer.com/article/10.1007%2Fs11625–011–0132–6#page–1 (Accessed on 8 July 2019).
359. At Aetna, a CEO's Management by Mantra, the New York Times, (2015). http://www.nytimes.com/2015/03/01/business/at-aetna-a-ceos-management-by-mantra.html?_r=1 (Accessed on 8 July 2019).
360. Building the New Leader: Leadership Challenges of the Future Revealed, (2014). Hay Group. http://www.haygroup.com/downloads/MicroSites/L2030/Hay_Group_Leadership_2030%20whitepaper_2014.pdf (Accessed on 8 July 2019).
361. Capitalizing on Complexity, (2010). IBM. http://www–935.ibm.com/services/us/ceo/ceostudy2010/ (Accessed on 8 July 2019).
362. David, J., Snowden, M. E., & Boone, A., (2007). Leaders Framework for Decision-Making, Harvard Business Review. https://hbr.org/2007/11/a-leaders-framework-for-decision-making (Accessed on 8 July 2019).
363. David, R., & William, R. T., (2005). Seven Transformations of Leadership. Harvard Business Review. https://hbr.org/2005/04/seven-transformations-of-leadership (Accessed on 8 July 2019).
364. Defining a Profession: Core Competencies for Sustainability, (2013). The College of Estate Management. http://www.cem.ac.uk/about-us/our-research/reports-publications/defining-a-profession-core-competencies-for-sustainability.aspx (Accessed on 8 July 2019).
365. Developing the Global Leader of TomorrRidgesh ridge, ND http://www.unprme.org/resource-docs/developingthegloballeaderoftomorrowreport.pdf (Accessed on 8 July 2019).
366. Erik, F., & Kelli, L. L., (2011). Educating for sustainability: Competencies and practices for transformative action. Journal of Sustainability Education, http://www.jsedimensions.org/wordpress/

content/educating-for-sustainability-competencies-practices-for-transformative-action_2011_03/ (Accessed on 8 July 2019).

367. Habits of a Systems Thinker. Waters Foundation, http://watersfoundation.org/wp-content/uploads/2012/09/habits-onepage.pdf#page=1&view=FitH (Accessed on 8 July 2019).
368. Is Mindfulness the Next Step in Paving the Way Towards Capitalism 2.0?, (2013). http://ecoopportunity.net/2013/06/is-mindfulness-the-next-step-in-paving-the-way-towards-capitalism–2–0/ (Accessed on 8 July 2019).
369. Jeffrey, G., Mary, C., Gerard, S., & Mark, R., (2013). Leadership Character and Corporate Governance. Ivey Business School at Western University, http://iveybusinessjournal.com/topics/leadership/leadership-character-and-corporate-governance#.VJx6b14AAA (Accessed on 8 July 2019).
370. Leadership and the Corporate Sustainability Challenge: Mindsets in Action, (2009). Avastone Consulting. http://papers.ssrn.com/sol3/papers.cfm?abstract_id=1118071 (Accessed on 8 July 2019).
371. Leadership Competencies for Corporate Citizenship, Boston College, (2010). http://ccc.bc.edu/document/docWindow.cfm?fuseaction=document.viewDocument&documentid=1352&documentFormatId=2397 (Accessed on 8 July 2019).
372. Leadership in a Rapidly Changing World: How Business Leaders are Reframing Success, (2012). Ashridge, http://www.unprme.org/resource-docs/leadershipinarapidlychangingworld.pdf (Accessed on 8 July 2019).
373. Leadership Qualities and Management Competencies for Corporate Responsibility, (2006). Ashridge, https://www.ashridge.org.uk/Media (Accessed on 8 July 2019).
374. Leadership Styles that Drive Sustainability Korn/Ferry Institute, (2011). http://www.kornferryinstitute.com/sites/all/files/documents/briefings-magazine-download/Leadership%20styles%20that%20drive%20sustainability%20.pdf (Accessed on 8 July 2019).
375. Library/Ashridge/PDFs/Publications/LeadershipQualitiesAndManagementCompetenciesForCorporateResponsibility.pdf (Accessed on 8 July 2019).
376. MSU Global Sustainability Competencies, Michigan State University, ND. http://sustainabilityspecialization.msu.edu/competencies.pdf

(Accessed on 8 July 2019).

377. People matter: Enabling leadership for sustainability, world business council for sustainable development, ND http://www.wbcsd.org/Pages/EDocument/EDocumentDetails.aspx?ID=13604&NoSearchContextKey=tru e (Accessed on 8 July 2019).
378. Personal Capacities for Authentic Leaders, (2012). http://emergentbydesign.com/2012/01/29/9-personal-capacities-of-authentic-leaders/ (Accessed on 8 July 2019).
379. Peter, S., Bryan, S., Nina, K., Joe, L., & Sara, S., (2008). The Necessary Revolution: How Individuals and Organizations are Working Together to Create a Sustainable World, 48–61.
380. Recruiting CSR competent leaders: Six criteria for CEO succession planning and recruitment, (2015). Strandberg Consulting, http://corostrandberg.com/publications/corporate-social-responsibility/recruiting-csr-competent-leaders (Accessed on 8 July 2019).
381. Social Innovation from the Inside Out, Stanford Social Innovation Review, (2014). Warren Nilsson and Tana Paddock, http://www.ssireview.org/articles/entry/social_innovation_from_the_inside_out (Accessed on 8 July 2019).
382. Susan, C., Schneider, M. Z., & Ramesh, M., (2010). Developing Socially Responsible Behavior in Managers: Experimental Evidence of the Effectiveness of Different Approaches to Management Education. http://www.ingentaconnect.com/content/glbj/jcc/2010/00002010/00000039/art00004 (Accessed on 8 July 2019).
383. Sustainability and Leadership Competencies for Business Leaders, Business for Social Responsibility, (2012). http://www.bsr.org/reports/BSR_Sustainability_Leadership_Competencies.pdf (Accessed on 8 July 2019).
384. Sustainability Leader Competencies: A Grounded Theory Study, (2011). Pamela G. Schwalb, http://digitalcommons.unl.edu/cgi/viewcontent.cgi?article=1030&context=aglecdiss (Accessed on 8 July 2019).
385. Sustainability Leadership: Linking Theory and Practice, Wayne Visser and Polly Courtice, (2011). http://papers.ssrn.com/sol3/papers.cfm?abstract_id=1947221 (Accessed on 8 July 2019).
386. Systems Thinking, Center for Ecoli15, (2015). http://www.ecoliteracy.org/essays/systems-hinking (Accessed on 8 July 2019).
387. The Dawn of System Leadership, Stanford Social Innovation Review,

(2015). Peter Senge, Hal Hamilton, John Kania. http://www.ssireview.org/articles/entry/the_dawn_of_system_leadership (Accessed on 8 July 2019).

388. The Future of Leadership for Conscious Capitalism. Barrett C. Brown, ND https://associates.metaintegral.org/sites/default/files/MetaIntegral_Brown_The%20future%20of%20lea dership%20for%20conscious%20capitalism_20131018.pdf (Accessed on 8 July 2019).
389. Transforming Sustainability Education at UBC: Desired Student Attributes and Pathways for Implementation, (2013). http://sustain.ubc.ca/sites/sustain.ubc.ca/files/Sustainability%20Attributes%20March%202013.pdf (Accessed on 8 July 2019).
390. Triple Strength Leadership, Harvard Business Review, (2013). https://hbr.org/2013/09/triple-strength-leadership (Accessed on 8 July 2019).

INDEX